NEMASKET
RIVER HERRING

A HISTORY

Michael J. Maddigan

natural
HISTORY
PRESS

Published by Natural History Press
A Division of The History Press
Charleston, SC 29403
www.historypress.net

Front cover, bottom: A portion of Lake Assawompsett, spawning ground of Nemasket River herring, as seen from near Tamarack Park, Lakeville, Massachusetts. *Photograph by Michael J. Maddigan.*

All maps and plans by Michael J. Maddigan. Plans of the Star Mill fishery based on earlier plans by James F. Maddigan Jr.

First published 2014

Manufactured in the United States

ISBN 978.1.62619.662.9

Library of Congress CIP data applied for.

It cannot but affect our philosophy favorably to be reminded of these shoals of migratory fishes, of salmon, shad, alewives, marsh-bankers, and others, which penetrate up the innumerable rivers of our coast in the spring, even to the interior lakes, their scales gleaming in the sun; and again, of the fry which in still greater numbers wend their way downward to the sea.

−Henry David Thoreau,
A Week on the Concord and Merrimack Rivers *(1849)*

And there shall be a very great multitude of fish…

−Ezekiel 47:9

CONTENTS

ACKNOWLEDGEMENTS

In 1849, when Henry David Thoreau wrote of "dim visions…of miraculous draughts of fishes, and heaps uncountable by the riverside," he also asked "to know more of that race, now extinct, whose seines lie rotting in the garrets of their children." Helping me to know more about the Nemasket River and its own extinct race of fishermen were several people whom I wish to sincerely thank, namely Tabitha Dulla, commissioning editor at The History Press, for her unfailingly sound advice and support throughout this project; Ryan Finn, project editor at The History Press, for invaluable help with the manuscript; the production and marketing teams at The History Press for the remarkable job they do in popularizing local history; Dave Cavanaugh and Bill Orphan of the Middleborough-Lakeville Herring Fisheries Commission for proofreading the manuscript and offering their insights, as well as providing revised population estimates for the period between 1996 and 2013; Danielle Bowker and the Middleborough Public Library for generously permitting me to reprint several images from the library's extensive collection; and the staff of the Taunton Public Library for their helpfulness with locating information and maps relating to the herring at East Taunton. All have contributed to bringing alive "tales of our seniors sent on horseback…perched on saddle-bags, with instructions to get the one bag filled with shad, the other with alewives."

INTRODUCTION

A curious phenomenon happens each March and April in the Nemasket River in southeastern Massachusetts. A solitary silver fish makes its way upstream all the way from Mount Hope Bay, its shiny body glinting in river water seemingly colored bottle-brown by tannins and iron ore. Soon that single fish is joined by dozens of others, alewives and bluebacks alike, then a torrent, literally hundreds of thousands, of river herring fighting their way upstream against the current, blackening the river with their vast multitude. Crowding, thrashing, leaping, the water is almost boiling. Herring gulls, silver and white, wheel gracefully overhead, and the cry goes out, "Herring have come!"—announcing to all the return of the river herring and with it the arrival of yet another spring.

Most of us who grew up in southeastern Massachusetts towns like Middleborough can vividly recall the herring runs of early spring, when thousands of alewives and bluebacks would make their tortuous way upstream to spawning grounds in sandy-bedded ponds. I can still remember going as a child to the local fishladder with my grandfather and brother, net in hand, and watching the fleet flashes of silver as the river herring darted by under the waters before making their way heroically through the ladder. Efforts to catch the fish barehanded proved futile, but a quick dip of the net into the river would result in a heavy load of flopping fish, one of which we would but momentarily hold (all slippery and thrashing as it was, jagged belly and twisting tail) before releasing it back into the river to continue its upstream journey.

Today, the Nemasket River in Middleborough and neighboring Lakeville supports the most populous river herring run in Massachusetts, one of the largest in New England and one that is regarded as among the most critical on the eastern seaboard, with just over 1 million fish having passed upstream to spawn in 2002. Signs once proclaimed the Nemasket run as the largest in the world. Although the size of the local run is attributable to the fact that the river drains the greatest expanse of naturally occurring fresh water in Massachusetts, nearly six thousand acres of which form an ideal spawning ground for the fish, it is equally the product of the historical development of the Nemasket River fishery.

Of all non-domesticated animals, it is arguably the river herring that has had the greatest impact on the inland communities of southeastern Massachusetts, including Middleborough. No other fish, no mammal and no bird has had as profound or lasting an influence on the inhabitants of the region as has the river herring. A report on the history and management of Massachusetts herring published by the Massachusetts Division of Fisheries & Wildlife agrees, labeling the the fish the "most symbolic of the New England fishing tradition." (Sorry, sacred cod).

For thousands of years, local inhabitants have interacted with the herring. The herring fishery is the first subject touched on in Mertie Romaine's comprehensive *History of the Town of Middleboro*, and the fish figures prominently on the seal of the Middleborough Historical Commission, two indications of its relative importance in the historical development of the town.

What follows is a cultural ecology that reveals a story of how a single fish, the herring, romantically and somewhat improbably shaped the culture and society of a small New England town—Middleborough, Massachusetts—a community once known to Native Americans, appropriately enough, as *Nemasket*, "place of fish."

HERRING AND
THE NEMASKET

NATURAL HISTORY OF THE HERRING

The subject of this work, the Nemasket river herring, is in fact two fish: the alewife (*Alosa pseudoharengus*) and the closely related blueback (*Alosa aestivalis*), both members of the herring family (*Clupeidae*). Although alewives and bluebacks have been mistaken for one another since their first encounters with the English and are today managed as a single species by the United States National Marine Fisheries Service, one distinguishing characteristic differentiating the two is the subtle color variation of the stripe running the length of the fishes' backs—greenish-gray on the alewife and blue on the blueback.

In 1885, Noah Hammond of Mattapoisett noted the alleged difference between Taunton and Mattapoisett River herring and their counterparts in the Agawam River in Wareham:

> Some people find it difficult to detect any difference between Taunton and Mattapoisett river herring and the Wareham variety except by the size. There is, however, a very marked difference in the color of the skin under the scales. That of the Taunton river is a sort of variegated orange and of the Wareham river a dark blue. I have never known this test to fail.

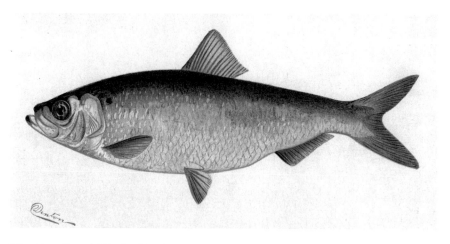

"Alewife or Branch Herring," by Sherman F. Denton, chromolithograph. *From* Fish and Game of the State of New York *(Forest, Fish and Game Commission, 1901).*

Chromolithograph by *National Geographic* artist and naturalist Hashime Murayama, 1930s. The deeper-bodied alewife appears at top, with two marine herring below. *Author's collection.*

The sea or herring gull is closely associated with herring. The bird's presence in early spring is indicative of the herring's arrival. *Author's collection.*

In fact, what Hammond may have been observing was the difference between alewives and bluebacks. The peritoneum or body cavity lining of the alewife is a very pale pearl-gray to delicate pale pink, while that of the blueback is charcoal. Other distinguishing features are the alewife's deeper belly and a large aqueous eye that is bigger than that of the blueback.

River herring are a migratory or anadromous fish, the latter term derived from the Greek for "up-running" in reference to the fact that the fish lives in salt water but swims upstream to spawn in fresh water. The Nemasket herring follow a three-year migratory cycle. Spawned and hatched in the fresh water of Lake Assawompsett and the ponds above it in southeastern Massachusetts, juvenile fish travel downstream to the sea, where they mature, with adult specimens reaching fifteen inches in length. Three years later, these same fish return as adults to their spawning grounds, with the spawning migration commencing as early as late March and continuing through mid-May.

River herring reach sexual maturity at an age between three and five years, whereupon they return to spawn in the same waters in which they were hatched. Noah Hammond remarked on this phenomenon in an 1885 interview in the *Boston Globe*: "A peculiarity of herrings is that they never mistake their stream. For example, the herrings that enter Wareham river are of two distinct species, but, upon reaching the point where the Agawam river forks, they separate and are never found mixed above the fork in either river."

Herring spawn in a variety of waters, including estuaries, broad rivers and slow-moving streams but most commonly are seen in ponds, including those with sand or gravel bottoms and underwater vegetation. There, mature herring broadcast 60,000 to 100,000 eggs and milt concurrently over the surface rather than in a nest on the lake bed. Although the number of eggs laid is enormous, typically less than 1 percent survive. The demersal pinkish-hued eggs sink to the bottom of the pond, temporarily clinging to whatever they come into contact with. Immediately after spawning, adult herring return to the sea, and no particular care for either eggs or juvenile fish is provided.

Transparent herring larvae measuring about a quarter inch long hatch within several days and soon begin feeding on zooplankton such as *Daphnia*. Herring fry remain in their spawning grounds throughout summer before returning downstream to more saline waters with the onset of autumn.

A social fish, river herring remain in proximity to their natal streams upon returning to the sea, congregating in schools numbering in the thousands. Incidences of seines on the lower Taunton River pulling in that many of the fish in a single haul were frequently noted during the nineteenth century.

Herring spend three to four years in the ocean before returning to their native pond to spawn and may spawn up to four times in their lifetime. Herring may live up to ten years, although the expected lifespan is a mere five hours, so high is the mortality rate of juvenile fish, with less than 1 percent surviving to make the fall migration to the sea.

As part of the local ecological system, river herring are an important link in the riverine food chain. Herring are planktivores, feeding mainly on zooplankton, although other juvenile fish and fish eggs constitute a part of the diet of the mature fish. Conversely, herring—and, in particular, herring fry—are subject to predation by other species, including American eels, striped bass, yellow perch, lake trout, white perch and cod. Additionally, freshly spawned herring eggs are readily consumed by perch. Other predators include harbor seals, river otters and mink, as well as piscivorous birds such as great blue herons, ospreys, bald eagles, cormorants and sea gulls. River herring are particularly associated with this last bird. The herring gull (*Larus argentatus*) is a scavenger. Although noted for great acrobatic ability in the air, it is skilled at neither diving nor swimming under water. It relies for its food, in part, on feeding on weak, injured, dead or dying fish, including herring. Consequently, it finds its food "where many fish are crowded together in a limited space in a net or weir or in a narrow, shallow stream," such as a fish ladder.

What's in a Name?

Although river herring and marine herring are distinct, because many early English settlers mistook them for their marine counterparts, alewives and bluebacks collectively came to be known as "herring." (In fact, the alewife's Latin name, *pseudoharengus*, means "false herring.") Mary Hall Leonard, writing at the start of the twentieth century, remarked on this etymology, noting that in southeastern Massachusetts "parlance…the name, 'alewive' was long ago dropped for 'herring.'" Further confusing the naming issue is also the fact that, historically, the term "alewife" has at times been used synonymously with "river herring," at which time it was understood to pertain to the blueback as well.

Regardless of its name by which they are known, river herring were recognized as distinct from marine herring from the earliest days of English settlement by perceptive observers. Both Roger Williams and Charles

"Shad, Alewife, Herring." *From* Harper's Magazine, *May 1880.*

Whitborne described the alewife as "much like a herring," while John Josselyn's early description of the flora and fauna of New England lists river herring and marine herring separately. One of the features distinguishing river herring from marine herring is their deeper body, a characteristic attributed by some as the origin of the alewife's English name, as female tavern keepers (or ale-wives) were stereotypically stout, buxom women. In 1675, Josellyn wrote, "The Alewife is like a herrin', but has a bigger bellie therefore called an alewife." In truth, the name's provenance is more ambiguous. Sources contemporary with and earlier than Josselyn (as well as Josselyn himself) refer to the fish as "allizes," "alooses" and "aloof," terms that drew on earlier source words in English (*allowes* and *alooses*), French (*aloes*) and Latin (*alausa*). It is likely that it was from these words that the similar-sounding term "alewife" was derived. For their part, the Wampanoag natives of southeastern Massachusetts knew the herring by at least two names, one of which was *aumsuog*, meaning "small fishes."

Besides its deeper belly, the river herring had another noteworthy distinguishing characteristic in the scutes or sharp serrations on the midline of its belly, a feature absent from sea herring and one that gave the alewife yet another of its names: "saw belly." Because of this notable difference, it has been said that a practiced hand can easily distinguish between the alewife and the sea herring by touch alone.

The Assawompsett Pond Complex and the Nemasket and Taunton Great Rivers

Nemasket alewives and bluebacks have an ideal spawning and nursery habitat. The Nemasket River's headwaters encompass nearly six thousand acres, including the largest natural bodies of fresh water in southeastern New England: Lake Assawompsett, Pocksha Pond, Long Pond, Great Quitticus Pond and Little Quitticus Pond. Located in the towns of Middleborough, Lakeville, Freetown and Rochester, these five interconnected ponds are today known as the Assawompsett Pond Complex (APC), and they are used to supply the cities of New Bedford and Taunton with water. With their fine sandy bottoms, great surficial expanse and clear, clean water, these glacially created ponds not only provide perfect spawning grounds for herring, helping to account for the success and the size of the Nemasket run, but also host a variety of fish in addition to herring, including largemouth bass, bluegill, chain pickerel, yellow perch, white perch, pumpkinseed, white sucker, brown bullhead, golden shiner, tessellated darter, lake chubsucker, black crappie and northern pike.

Receiving the waters of these ponds is the Nemasket River, which threads nearly twelve miles northward through areas of woodland, palustrine wetland and gravel-filled hills to its confluence with the Taunton River. From its headwaters at Assawompsett to the Taunton River, the Nemasket falls only thirty-eight feet and is in consequence a slow-moving, winding stream with stretches of level terrain and wetlands fringing the river.

Perhaps appropriately for a river described as "Great," the Taunton River, with which the Nemasket is intimately connected, has no single source and is instead formed by the confluence of the Matfield (Ahquannissowamsoo) and Town (Nunckatesset) Rivers at Bridgewater. Historically, the Taunton was known to the native population as Titicut and later to the English as Titicut Great River and Taunton Great River.

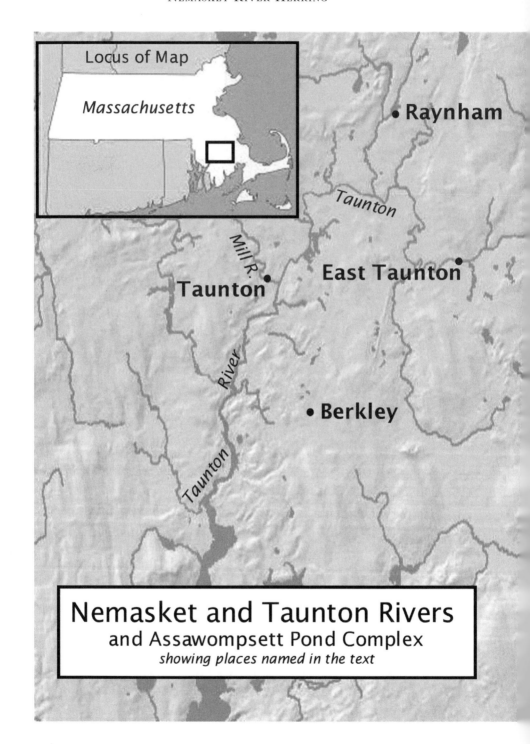

Locus of Map

Massachusetts

Nemasket and Taunton Rivers
and Assawompsett Pond Complex
showing places named in the text

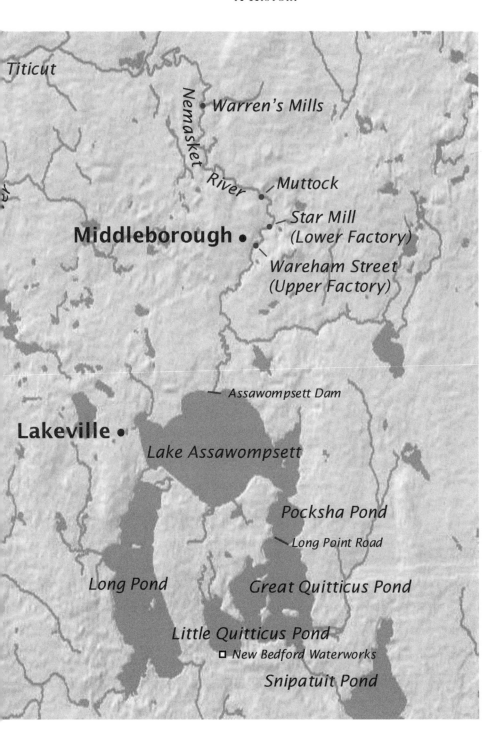

Titicut

Nemasket River

Warren's Mills

Muttock

Star Mill
(Lower Factory)

Middleborough •

Wareham Street
(Upper Factory)

Assawompsett Dam

Lakeville •

Lake Assawompsett

Pocksha Pond

Long Point Road

Long Pond

Great Quitticus Pond

Little Quitticus Pond

□ New Bedford Waterworks

Snipatuit Pond

Lake Assawompsett, the largest natural body of water in Massachusetts and an important spawning ground for herring, 1910s. *Author's collection.*

Taunton River at North Middleborough. Photograph by A.M. Hinley, circa 1901. The Taunton remains New England's longest free-flowing river. *Author's collection.*

From its origins in Bridgewater, the Taunton flows through level landscape, draining the commonwealth's second-largest watershed of 562 square miles and flowing through Bridgewater, Halifax, Middleborough, Raynham, Berkley, Dighton, Freetown, Somerset and Fall River. The river drops only twenty-one feet in the forty miles from its headwaters to its confluence with the Quequechan River at Fall River, resulting in a meandering course and permitting tidal influence to reach as far upstream as East Taunton, eighteen miles from the sea. Between Bridgewater and Fall River, the Taunton River receives the flow of a number of tributaries, most notably the Winnetuxet, Nemasket, Cotley, Forge, Mill, Three Mile, Segregansett and Assonet Rivers. Below its confluence with the Three Mile River at Dighton, the Taunton broadens into a wide, brackish tidal estuary before finally emptying into Mount Hope Bay at Fall River, a place Sidney Lanier poetically described as "where Taunton helps the sea." The Taunton River was a vital transportation link during the pre-contact and contact eras, and native occupation of the river's watershed dates back some twelve thousand years. Long recognized has been the river's historic influence within the region, supporting shipbuilding, maritime commerce and industry.

Historical Abundance of Nemasket Herring

The historical abundance of herring in the Nemasket and Taunton Rivers is an aspect of the region's natural history dating back millenia. As opposed to other facets of the Nemasket's history, however, records concerning the Nemasket herring are scant, both relative to the timing of their run and the estimated number of fish running. Much of the evidence remains anecdotal, such as Thomas Weston's comment that "a person wading into the river, with a bushel basket, could in some seasons dip up a basket full of these fish." Novelist Joseph C. Lincoln would later provide an eloquent analogy demonstrating the multitudes of fish passing up the region's rivers: "Spread the fingers of your hand as far as you can spread them and let each finger represent the back of a fish. They were as close together as that."

Slightly more concrete are reports in local newspapers, like the *Middleboro Gazette* and *Old Colony Memorial*, dating from the 1850s and later that documented the total catch from each of the weirs operating in Middleborough: Muttock (now Oliver Mill Park), the Lower Factory (Star

The Nemasket River was once described as so "thick" with herring that one could walk across on the fishes' backs. *Photograph by Michael J. Maddigan.*

Mill) near East Main Street and the Upper Factory at Wareham Street. Sadly, more official records only too rarely recorded such information for posterity. In some years, such as 1871, 1873, 1874 and 1875, it seems that no records were kept at all. Other information must be culled from old newspaper accounts that documented either the paucity of each year's harvest or its abundance.

1854

The catch in this year exceeded that of the previous two or three years. "Somedays 15,000 to 20,000 have been taken." Although the *Namasket Gazette* reported that some 200,000 herring had been caught during the 1854 season, the commonwealth reported a far higher catch: 350,000, valued at $1,800.

1855

The first herring of the 1855 season were noted at noon on April 9. "Some three or four thousand were taken at the Muttock works during the afternoon…. They are thought to be of a larger size than usual." On Monday, April 23, 1855, 24,000 fish were taken, 23,400 the following day and 29,300 on Wednesday (of which 20,000 were taken at the Upper Factory), making a total of 96,700 within a week's time. "Somedays they have been very abundant, 75 or 100 being taken at a single scoop of the net."

1856
April 28: sixteen thousand taken at Muttock; April 29: twenty-two thousand taken at Muttock; April 30: twelve thousand taken at Muttock. Despite these numbers, the *Namasket Gazette* asserted that "the prospect is not good for a large supply of this kind of fish, the present season."

1857
"About two hundred and twenty-five thousand herrings have been taken at the weirs in this town the present season, valued at $1125. Forty thousand a day are sometimes taken when there is an extra run."

1860
73,768 taken.

1862
Undoubtedly a banner year. The *Taunton Gazette* reported:

> There has been nothing like the quantity of herrings in Taunton river, during any year for the last fifteen years. They are literally too abundant. We hear that in the seine near East Taunton as many as sixty thousand have been taken at one sweep; and on one occasion within a week the owner of the privilege was compelled to load the herrings into his carts and use them for the purpose of enriching his land.
>
> Several thousand were thus disposed of. We do not learn that the catch of shad is much in excess of former years; but of herrings there is a surplus.

1864
The first herring of the season were taken at "Namasket Weir" (Lower Factory) on April 19, with three thousand taken that day.

1865
"Over 42,000 were taken at the lower weir, in three days in April."

1867
A total of 100,713 fish were taken (50,757 at lower weir; 49,956 at upper weir).

1868
The first herring of the season was caught at the weir by Joseph Bisbee during the third week of April. Total taken: 74,206 (36,419 at lower weir; 30,019 at upper weir).

1869
Total taken: 77,315 (54,375 at lower weir; 22,940 at upper weir).

1870
"Herring Market—Caught at Star Mills privilege this week about 20,000. Condition good, being both healthy and fat; price $1.25 to $2 per hundred. It is said that the natives of old Middleborough already begin to look more cheerful and better every way." Total taken: 84,724 (37,640 at lower weir; 47,084 at upper weir).

1872
The herring reached Middleborough on April 15. Total taken: 111,848 (52,553 at lower weir; 59,295 at upper weir).

1873
Herring were reported at East Taunton in mid-April. Later, in mid-May, "We wish some of the incredulous ones could have been with us Tuesday afternoon at the fishway at Sherman's Shovel Works and seen some of the thousands of herrings winding their way toward Assawampsett…. There were thousands following thousands up the devious way."

1875
During the third week of April, it was reported that "herring are hurrying up the intricate way of the Nemasket river and the men at the weirs are trapping them in all ways possible." In September 1875, it was further reported that "young herring by the million have been going down the rivers from Middleboro' to the sea."

1876
"On account of the cold weather continuing so late…the herrings were very late in reaching here. The first ones taken at the Star Mills was on the

14th of April and of the 47,543 taken at that weir, only 5,000 were taken up to the 8th of May…. Also quite a large number were lost, as they could not be saved on account of hot weather during the last of the fishing season." Total taken: 78,068.

1877
Four hundred fish were caught on the third Monday of the month, which was the first day of fishing at the Star Mill weir.

1878
Total taken: 166,328.

1879
Total taken: 81,213.

1880
Total taken: 70,160.

1881
Total taken: 87,820.

1882
Total taken: 50,100.

1883
Total taken: 139,153.

1884
Total taken: 44,515, "which is the smallest number of fish caught in any year of which we find a record." "The catch of alewives at Middleboro has been but about one-third of the usual average, and at East Taunton the fishery is considered a failure. The herring at the latter place were kept out of the river by large shoals of menhaden."

1885
"Only 35,000 herring have been caught at the Nemasket and Star Mills dam in Middleboro this season; and the purchaser of the privilege for $199 is thought to be 'pinched' about $100 worth." Total taken: 35,650.

1886
"Over 150,000 herrings have been taken at the weir in Middleboro this season."

1887
Total taken: 201,635. The number taken from the Taunton River "was so great this season that prices are lower now than they have ever been in the history of the industry."

1889
The 1889 catch of 134,000 was described as the heaviest in twelve years. Randall Hathaway, the purchaser, paid "$345 for it and cleared $100 on the operation."

1893
About nine hundred barrels were taken.

1894
"225,000 were gathered in, which was an exceptionally large number."

1895
"The herring season is over and the catch foots up 135,000."

1897
Although it was reported that "there were thousands of them near the Middleboro fishways the first of the [first] week" of April, "the catch of herring has been small thus far, owing to the high water and cold water. The fish are noticeably larger than usual this season." About four hundred barrels were taken.

1898
"A few herring are making their appearance at the Nemasket fish ways but the continued cold weather has made the fish coy in coming up stream." "The herring catch will be wound up this week and it has been light. About 225 barrels have been taken."

1899
"The first herring of the season came up the river Tuesday [in early April], L. Deane catching 35 at the Muttock dam.... Nemasket river was alive with herring Tuesday [the last week of April], the first warm day. About 10,000 were captured that day."

1900

"It is reported that a few herring have been seen at the Muttock weir...[By late April, it was reported that] herring are arriving very plentifully now, 40 or 50 barrels being caught daily at the Star mills and Muttock."

1901

The run during the second week of May was the largest of the season to date, "the warm days bringing the fish upstream in great numbers."

1902

"The first herring of the season invaded the Nemasket waters [the first week of April]. They are a little late but none the less welcome." The season peaked during the final week of April, when about ten thousand were being reported taken at the Star Mill on the permitted days.

1903

The fish arrived "considerably earlier than usual," being noticed at the Star Mill the third week of March. The catch in 1903 was particularly light, owing to the cold winter and the high water levels, the largest single-day catch being but one hundred fish. "The herring catch has practically ended. It has been light this year, owing to the high water prevailing in the river and there has been small demand for the fish as bait." "One comfort is that the demand for fish has not been very brisk."

1904

"The herring fishery in town will not be a brilliant success this year. At the [Star] mills fishway from 4,000 to 5,000 alewives have been taken daily, but there is no market for them."

1905

Although it was stated in May that "the fishing was reported better this year than for some years past," the initial report had noted the arrival of the alewives in smaller numbers. By April 28, Randall Hathaway had "taken in 26,000 herrings out of the wetness thus far this week." James Creedon labeled the 1905 season as "one of the most successful for years," although this was a judgment based on the price the fish fetched rather than the number of fish running: "the ruling price for the fish early in the year was very high, and a good profit was realized. The run was fairly good."

1906

Walter Blair, purchaser of the Nemasket privileges, "has made some fine catches this week, netting 35 barrels, Tuesday [May 1]. Until the warm days of the first of the week the catch had been rather light, but the fish have since been coming up the river in greater numbers."

1907

Walter D. Blair, once more lessee of the Nemasket privileges, caught the first herring of the season at the Star Mill on April 22. "Walter Blair is getting some good catches of herring just now at the Star Mill weir, the run of late this week having been especially good."

1908

The 1908 run was reported as being several days later than that of 1907. Nonetheless, herring were reported in large numbers with good catches at the Star Mill, and seventy-seven barrels were reported shipped through April 30. "The catches thus far have been heavier than for years. One day the catch was about 60 barrels, and on others it was correspondingly good." By the end of the season, the fish were being sold for bait in Boston.

1909

Despite the fact that there were said to be a large number of fish ("there is an exceptionally large run of herring and alewives in the Nemasket river. The fish have not been seen in such great numbers for many years.")—a circumstance attributable to the breakdown of the Wareham Street dam three years earlier in 1906, which permitted the upstream passage of large numbers of alewives—cold weather and frequent rain were blamed for delaying and hindering the progress of the fish in 1909. Walter D. Blair, who leased the privilege that year, reported his biggest catch of fifteen barrels on May 5.

1911

The run during this year was stated to be "one of the best in recent years."

1914

"The first herring of the season came up the river last week [mid-April] and about 600 were caught at the Star Mills weir."

Herring harvest at the Star Mill. Photograph by George Morse, spring of 1910. *Author's collection.*

1915

"The first catch was Monday [April 19] when 15 barrels were taken at the Star Mill weir." On May 28, it was reported that "the alewife fishing for the season is practically over and the catch has been smaller than usual, about 225 barrels having been taken."

1916

"Herring put in their appearance" during the final week of April, "but only a few went through the fishway. The warm weather the first of the week gave them a start and they passed up in great numbers Tuesday and Wednesday," May 2 and 3.

1917

This year, the fish were taken by the town as a wartime measure. As of May 4, the fishermen were "ready for the fish whenever the weather becomes warm enough for them to come up the river. They have secured but few yet."

Monday and Tuesday, May 14 and 15, were the first days that a "catch of any number of herring was made, about 1,200 being netted."

1918
The run was described as "smaller than usual."

1920
"Herring are unusually plenty in the Nemasket River this year and more than the average number have passed through the run at Wareham street."

1940
The run reached its height at the start of May, and while lessee James Ferreira claimed that a haul of 10,000 barrels was possible, a depressed market encouraged him to seek only 3,500.

1943
Fifteen thousand barrels taken. This was the largest recorded harvest ever taken from the river.

1944
The year of the die off, when the run abruptly ended on May 4. Prior to that, the fish were numerous. "The river was seething with millions of the fish as [the first week of May] began. They were thick in the fishing pool below the dam, they were forcing their way not only up the board sluice in the dam, provided for a fishway, but hurling themselves up through the boiling water at two other openings, and the pools of the cement fishway at Wareham street were packed almost to capacity." Up through May 1, an estimated five thousand barrels had been taken, and three thousand more were anticipated.

1946
Nine hundred barrels.

1947
"The run proved to be a poor one," with the buyer taking only one hundred barrels. The lack of fish should not have been surprising, given the massive die off in 1944.

1948

The total catch was about two hundred barrels, and the run "proved to be the worst in a long time."

1949

"The river was filled almost solid with the fish."

1950

"The fish were in the stream packed body to body, it would seem."

1952

The first two weeks of the run in April yielded about four thousand barrels. "One dip of the net is reported to have yielded 60 barrels." Although only six thousand barrels were ultimately taken, the run was described as "profitable."

1953

About 337 tons, or 1,700 barrels, were taken.

1954

The run for this year was described as "short" and "unprofitable," with only about nine hundred barrels being taken. New Bedford's sandbag dam at Assawompsett was blamed for the poor return.

1958

Middleborough selectman Rhodolphus P. Alger stated that he had never seen a run as heavy as this year. "Herring gathered in the

Star Mill harvest, 1950s. The historic bounty of Nemasket River herring is evident in this photo. *Middleborough Public Library.*

Seining at the Star Mill, 1950s. By this time, seining had replaced the more laborious process of hand-netting Nemasket River herring. *Middleborough Public Library.*

fishing pool in such large numbers that the Nemasket River appeared black." Nearly five thousand barrels were taken.

1963
Commercial fishing began on April 1.

1964
On April 13, commercial fishing commenced, and a netful equalling nearly two hundred barrels was drawn.

1965
"A few herring were seen at Oliver Mill Park [Muttock]" on April 10, but the following day, they were outnumbered by spectators. The run was described as being near peak the final week of April.

1966
"The herring run began to build up to peak volume over the week end" during the second week of April.

1996
An estimated 696,666 herring passed upstream.

1998

The run was estimated at 651,441.

1999

The run was estimated at 766,694.

2000

Estimates of the number of fish passing upstream placed the number of herring at 560,986.

2001

The run was estimated at 284,498.

2002

Although one estimate placed the Nemasket alewife population at 1,361,691, the Atlantic States Marine Fisheries Commission (ASMFC) stated that the number observed was even higher—1,919,000.

2003

The run was estimated at 548,835.

2004

The run of 244,832 this year was half that of the previous year.

2005

Although the Nemasket herring population had been declining since 2002, this year saw a drastic 80 percent drop in river herring populations throughout the region. Herring failed to "appear in appreciable numbers until early April, rather later than usual. By the second week of May there were very few fish to be seen…. 33,964 herring were taken below the Wareham Street fish ladder, about half the number taken in 2004." The final estimate was 225,904. In response to the decline, the Massachusetts Division of Marine Fisheries Advisory Commission placed a three-year moratorium on the harvest, sale and possession of river herring throughout the commonwealth. Although the Nemasket as a regionally controlled river was exempt, the Middleborough-Lakeville Herring Fisheries Commission adopted the ban.

2006

"Herring appeared in the Nemasket in early March. There was a heavy push during early and mid April, but the push ended as fast as it came." The run was estimated at 313,242.

2007

"Herring appeared in the Nemasket in mid March this year. Although there were a few days of plentiful fish, the overall run was again bleak." The run was estimated at 462,000.

2008

Although it was stated at the time that "between 680,800 and 1,016,800 herring were estimated to have passed through the Wareham Street ladder, current estimates now place the year's population at only 392,451. Massachusetts Marine Fisheries extended the river herring moratorium an additional three years in the absence of significant population increases throughout the commonwealth. It was similarly extended by the local herring fisheries commission.

2009

The Nemasket run was estimated at 383,338.

2010

An estimated 489,931 fish passed through the Wareham Street ladder during the spring run.

2011

The commonwealth's moratorium on the harvest, sale and possession of river herring was extended indefinitely. The year's run was estimated at 512,139, the highest in eight years.

2012

Estimated population: 567,952.

2013

Estimated population: 840,033, "based on a continuous series of hand counts taken at the Wareham Street fish ladder between March 14 and May 26."

THE EARLY HISTORY OF THE NEMASKET HERRING

NATIVE AMERICANS AND HERRING

Following climatic and environmental changes that came in the wake of the retreat of the Wisconsin glacier about fourteen thousand years ago, the natural abundance of herring at Nemasket likely attracted the earliest inhabitants to the region: bands of Paleo-Indian hunters and gatherers, who are known to have been present as early as nine thousand years ago on the northern shore of Lake Assawompsett at Wapanucket, just east of the Nemasket River. There, the river provided a bounty of fish for food, including anadromous species, the high fat content of which provided an important nutritional supplement in winter.

During the contact and plantation periods (1500–1675), natives moved seasonally to be near food sources, relocating in spring to inland fishing grounds such as Nemasket and Muttock along the Nemasket River, Titicut on the Taunton River and Cohannet on the Mill River in Taunton.

Herring were most frequently captured by natives at these sites by netting, a practice documented by John Josellyn: "The alewives they take with nets like a pursenet put upon a round hoop'd stick with a handle in fresh ponds where they come to spawn." Additionally, larger nets made of natural fibers such as milkweed and dogbane and weighted with stone plummets may have been strung across rivers to trap fish. Vigils were kept beside the nets both day and night during the seasonal runs, and once captured, herring were skewered with "arrows, or sharp sticks."

To further facilitate the capture of herring, natives three to four thousand years ago began erecting weirs—wicker-like structures resembling fences constructed of vertical sticks staked into the riverbed and interwoven with twigs and brush. More substantial and permanent weirs were also constructed of stones. Weirs took different shapes, specific to the needs of local natives and the nature of the river, but all worked on the principle of funneling fish through a narrow outlet, where they could be taken easily. The last surviving native weir in the vicinity of Middleborough is the V-shaped stone weir on the Satucket River in East Bridgewater, preserved in part by having been submerged beneath the impounded waters of a millpond created when a dam was constructed downstream from the weir in 1819.

Locally, natives established stone fish weirs on the Taunton River at Titicut in North Middleborough, along the Nemasket River at Muttock and upstream from the Wading Place at East Main Street. The native villages that sprang up around these weirs became known as Titicut, Muttock and Nemasket, respectively, and each also featured a "wading place" where paths forded the river.

Although the Titicut weir had long since disappeared by 1859, memories of it persisted. In that year, Stillman Pratt, editor of the *Namasket Gazette*, visited the weir's site, which was located on the farm then owned by Albert G. Pratt on Vernon Street at Pratt's Bridge. "His extensive and beautiful farm embraces what…was called the old Indian Weir," noted Pratt. The weir reputedly stood near an enormous elm tree, "to which the Indians fastened their canoes and under the shade of which they laid their [catch]."

The capture of fish was not a solitary occupation but rather was engaged in by the whole community. Roger Williams noted of the Narragansetts, "with friendly joining they…stop and kill fish in the Rivers." Emery's *History of Taunton* described the communal Wampanoag practice of catching herring at the falls in the Mill River at nearby Cohannet, today's Taunton, noting that "hundreds of Indians would come from Mount Hope and other places every year in April, with great dancing and shouting to catch fish at Cohannit and set up their tents about that place until the season for catching alewives was past and would load their backs with burdens of fish and load ye canoes to carry home for their supply for the rest of the year and a great part of the support of ye natives was from alewives." Daniel Gookin also noted the number of natives who flocked to Wamesit, present-day Tewksbury, near the

The Taunton River at North Middleborough, early 1900s. Located near this site was the Titicut weir, where herring were captured. *Author's collection.*

Merrimack and Concord Rivers, where "there is a confluence of Indians, that usually resort to this place in the fishing seasons."

These gatherings naturally lent themselves as forums for English preachers eager to Christianize natives. On May 5, 1674, at the height of the alewife run, Gookin, accompanied by the "Apostle to the Indians" John Eliot, traveled to Wamesit to preach to the natives there. Closer to Middleborough, others may have done the same. John Cotton of Plymouth told Gookin that he had preached to natives at Nemasket, "wither come the praying Indians" of Assawompsett Neck and Titicut. Cotton confessed to Gookin that he took advantage of periods when the court was in session at Plymouth to preach, as "there are usually great multitudes of Indians from all parts of the colony," so it may not be unreasonable to assume that he did the same at other large gatherings of natives, such as the annual herring run.

Since native weirs permitted the capture of more fish than could be readily consumed, local Namaskets resorted to a combined technique of sun-drying and smoking to preserve their herring—the use of salt as a preservative was unknown during the pre-contact period. Although the fish may have been partially sun-dried, their oily nature would have necessitated smoking—otherwise they simply would have putrefied in the

Nemasket River, circa 1910. *Author's collection.*

sun. Fish were smoked over outdoor fires, although apparently in humid weather they were brought indoors for processing.

Herring constituted a portion of the native diet principally in the spring and was succeeded by non-spawning fish such as bass in the summer and eels returning inland in September. This diet was not as unvaried as it may appear, as it was complemented with game and vegetables such as squash, pumpkins, beans and corn, the latter of which composed a substantial portion of the native diet. Dried and smoked herring were generally prepared by native women as a stew, cut into small pieces and boiled with roots and vegetables. Frequently, powdered acorns were added as a thickener. Sometimes the fish was added to the boiled maize that was eaten frequently.

In preparing the fish, little thought was given to the many needle-like bones. Gookin marveled at the ability of natives to consume herring untroubled: "They were not in danger of being choked with fish bones; but they are so dextrous to separate the bones from the fish in their eating thereof, that they are in no hazard."

One important impact of the native weirs is that the easy availability of herring and the time required to properly preserve large quantities of the fish encouraged native sedentism. The location of fish weirs near wading places was a common arrangement, and such sites including Nemasket, Muttock, Titicut, Cohannet and Satucket naturally developed as seasonal village complexes that determined the later economic geography of the English by becoming the loci of English settlement.

The English Encounter the Herring

The earliest encounters between the English and river herring were marked by the Englishmen's astonishment at the abundance of the fish. While it was cod that would eventually be considered sacred in Massachusetts, the region's inland waterways correspondingly abounded with herring in equally astounding numbers. As with the cod, the bounty of hering was regarded as a manifestation of God's munificence and benignity. The preamble to the first law drafted at Plymouth governing the herring fishery was explicit in the view that "God by his providence hath cast the fish called alewives or herrings in the midst of the place appointed for the town of Plymouth." This view was still commonly held in 1700, about when William Briggs Jr. wrote that herring seemed "to be a sort of fish appropriated by Divine Providence to Americans," noting that its presence had afforded even remote communities like Dunstable, Massachusetts, relief during harsh winters.

The earliest contact-era accounts of the Nemasket region made note of the prodigious number of herring swimming in its rivers and streams. As early as 1616, the annual herring run of New England was already a noted phenomenon. Captain Charles Whitborne wrote of English explorers familiar with the Massachusetts coast that "experience hath taught them at New Plymouth that in April there is a fish much like a herring that comes up into the small brooks to spawn, and when the water is not knee deep they will presse up through your hands, yea, thow you beat at them with cudgels, and in such abundance as is incredible."

In 1622, John Pory repeated this description nearly verbatim when writing of the herring migration up Town Brook in Plymouth to spawning grounds in Billington Sea, and his and Whitborne's characterization soon became the standard metaphor with which to describe the abundance of herring in New England's rivers and streams. So overused did the phrase about beating back herring become that it soon was considered cliché.

Comments on the presence of the fish in heavy numbers appeared throughout the early contact period. Reverend Higgeson, writing in 1629, noted tersely of New England that "here is abundance of herring." Josellyn was a little more explicit, noting that in his travels he encountered two men who had taken ten thousand of the fish in the course of two hours without the benefit of a weir or other means, except "a few stones to stop the passage of the River." Thomas Morton noted in his influential *New English Canaan* (1632), "Of herring there is a great store, fat and fair, and to my mind as good as any I have seen," while William Wood in *New England's Prospect* (1634) remarked that the herring came "in such multitudes as is almost incredible, pressing up such shallow waters as will scarce permit them to swim."

Not surprisingly, it was not long before early English settlers sought to exploit this wondrous resource.

THE COLONIAL
FISHERY, 1620–1775

HERRING AS FERTILIZER

From the earliest days of colonial settlement, herring were regarded as an important commodity, valued not only as a food source but also (and more importantly) as an indispensable resource for the promotion of agriculture. The Pilgrims' survival following 1621 was ultimately ascribed to herring.

The Wampanoags of southeastern Massachusetts had long engaged in the practice of manuring cornhills with herring drawn from the rivers, planting a single fish with a few kernels of corn in a hill, a practice known as spot fertilization. Not only did decomposing herring provide organic matter to the soil, but their calcium- and lime-rich bones also helped neutralize acidic New England soil.

The English were taught to adopt the practice of spot fertilization by Tisquantum, more familiarly known as Squanto, as indicated by Bradford in his history *Of Plimouth Plantation*. Edward Winslow, writing on December 21, 1621, recounted that the Pilgrims had adopted the practice at the start of their first season with astonishing results: "We set the last Spring some twentie Acres of Indian Corne, and sowed some six Acres of Barly & Pease, and according to the manner of the Indians, we manured our ground with Herings, or rather Shadds, which we have in great abundance, and take with great ease at our doores. Our Corne did prove well, & God be praysed, we had a good increase in Indian-Corne."

The practice of spot fertilization with herring, however, was confined to "where the ground is not very good, or hath beene long planted and worne out." Fertile land had no need for such enrichment. John Smith in Virginia knew of this treatment for New England's "overworn" fields, noting the New England practice of "sticking at every plant of corn, a herring or two; which cometh in that season in such abundance, they may take more than they know what to do with." Additionally, the practice was employed strictly for the propagation of corn, which required planting in the ground and could not be broadcast sown as with other English grains, such as wheat, rye, barley and oats.

The drawback of spot fertilization with herring was that decaying fish attracted predation by dogs and wolves. Consequently, fields had to be watched constantly for a period of about two weeks, until the fish had rotted. Children supplied with stones were frequently given the task of warding off animals that might unearth the corn in efforts to get at the herring. Towns such as Taunton, Rochester and Ipswich approved bylaws requiring the fettering of dogs during the corn-planting season in order to prevent disruption of fields.

Eventually, the task of harvesting herring for fertilizer appears to have devolved onto children as well, so easily could it be done. Writing in 1711, William Briggs Jr. indicated that "these fish may be catcht by the hands of children in their nets while the parents have y'r hands full of work in the busy time of the Spring to prepare for planting."

The causal relationship between herring and survival was readily apparent to Pilgrim observers, who understood the critical role of herring in creating a sustainable agricultural foundation from the poor sandy and acidic soils that blanketed the region. Whereas the earliest grain crops imported from England had failed, Indian corn fertilized with herring had permitted colonists to survive their first season at Plymouth, a lesson deeply impressed on their descendants who settled Middleborough.

Such an understanding was pervasive throughout Massachusetts and beyond. Watertown residents in 1632 erected a weir across the Charles River even though they lacked permission to do so. Motivating them to take such illegal action was their experience the previous year of "falling very short of corn the last year, *for want of fish.*" About a half century later, failure to protect Mill River herring in Taunton was cited as the reason for residents there having to import corn. "Some of Taunton have been forced to buy

Above: Plymouth County towns, including Pembroke, memorialized their debt to herring through communal histories, folk rhymes and media, including this early twentieth-century postcard. *Author's collection.*

Left: Post–Civil War children's novels like *Soldier Rigdale* (1899) inculcated into Middleborough's historical consciousness the role that natives had played in disseminating knowledge of fertilization with herring. *Author's collection.*

Indian corn every year since the fish were stopped, who while they fisht, they'r ground used to have plenty of corne." Similarly in Weymouth, from whose Back River "the fish are literally shoveled out," residents attributed their survival to the presence of herring, a view enshrined in a common nineteenth-century South Shore folk rhyme:

> *Cohasset for beauty,*
> *Hingham for pride;*
> *If not for its herring*
> *Weymouth had died.*

Later, the use of herring for its manurial properties entered the New England consciousness, and the story of its origins continued to be repeated down to the present day. Reference to Squanto's pivotal role seemed obligatory in the literature of the nineteenth and twentieth centuries, particularly immediately preceding the Pilgrim tercentenary in 1920; it was also mentioned in numerous novels, plays and children's books, all of which helped perpetuate the herring's role in securing the survival of the Plymouth Colony.

FISHING RIGHTS AND ENGLISH WEIRS

In colonial Massachusetts, enormous value was placed on protecting local herring fisheries—action predicated on the belief that without the fish, the colony would cease to exist. By 1633, Plymouth had enacted restrictions on taking the fish for any purpose other than agricultural and by no other people than residents of the town. The law, with its exclusivity clause reserving herring solely to town inhabitants, became a model for other communities, including Middleborough.

The English were careful to acquire for themselves rights to local herring fisheries, so convinced were they of the fish's role in fostering their survival. The earliest purchase agreements with the natives at Middleborough explicitly included "libertie to make use of the alewives." Infringement on native rights to take herring and the erection of mills on traditional herring streams were likely factors contributing to growing animosity between natives and English in the third quarter of the seventeenth century.

Once fishing rights were secured, the colonists moved to exploit their newly acquired resource by encouraging private enterprise. In 1639, the General Court at Plymouth permitted the construction of "a herring ware to take fish" on the Jones River as a common stock venture and authorized construction of additional Plymouth weirs at Morton's Hole, the Bluefish River, Eagle's Nest, Green Harbor, the Eel River "or any other creek" in the hopes of initiating a spate of beneficial weir building.

Like their counterparts at Plymouth, early residents of Middleborough proved keen to exploit the local fishery to its greatest advantage. On July 20, 1683, the town voted "to build a substantial Ware for the taking of their fish" near the Wading Place on East Main Street. It seems that the weir was not built at that time, for a subsequent town meeting in September 1684 called on John Nelson to "get either a man or men to sett down a good sufficient herring ware, near about where the bridge now stands." This fieldstone rubble weir in time became known as the "stone ware" and later as the "old weir" in recogntion of it as the earliest one constructed by the English on the Nemasket.

The 1680s Nemasket weir eventually assumed the role of official town weir, particularly following 1687, when the town ordered the removal of all other weirs along the river. In 1749, the Massachusetts legislature further strengthened this connection when it prohibited the taking of herring from the Nemasket River at any point other than the "old Stone Ware." (Not until 1764–65 was the town permitted to take herring at Muttock.)

In contrast to Plymouth nearly fifty years earlier, however, Middleborough strenuously discouraged individual enterprise in the herring fishery, regarded as too valuable a commodity to leave to private competition. Nonetheless, steadily increasing herring prices in the first quarter of the eighteenth century convinced some residents of the financial advantage of building private weirs. In February 1724, a group of men proposed constructing a weir at Purchade Neck in northern Middleborough. The town, however, looked unfavorably on this proposal, concerned in part about maintaining monopolistic control over the local fishery, and at a town meeting in February "voted…to prevent the said wear from being built." Through these means, the town, backed by a 1709–10 act of the state legislature, exerted its authority over the herring fishery, from which private enterprise continued to be excluded.

Prohibiting weirs along the Nemasket was designed as a protective measure, as unauthorized weir construction elsewhere had destroyed herring runs. Weirs along the Merrimack River in northeastern Massachusetts that once had "abounded with plenty of fish, which hath been of great advantage to the inhabitants of the several towns, near said

Merrimack River, circa 1900. Middleborough saw a cautionary tale in the experience of rivers like the Merrimack, where uncontrolled weir construction had destroyed herring runs. *Author's collection.*

river," had nearly destroyed the "abundance of fish" there by 1735. Later, mills and dams completed the process.

Middleborough residents were clearly concerned with maintaining a viable herring fishery, as the fish continued to be used to manure the soil. On March 29, 1706, the town "voted that if there be any man in the town that doth not plant any Indian corn, he shall have no turn of fish, and he that plants so little that he needeth not a whole load of fish for it, he shall have no more than for what he doth plant; in which proportion it is to be understood that he shall use but one fish to a hill." The following year, residents were permitted one herring for each hill of corn planted. If no corn was to be planted, no herring was distributed.

The Herring Market

From its earliest days, the Nemasket fishery operated as a municipal enterprise. The town engaged men to catch herring that were originally distributed free to all Middleborough residents. Later, a nominal fee was charged, although the fee was waived for widows, spinsters and others without means. While the amount permitted under the fee system was initially a "load," eventually the number of fish allowed was decreased over time to just two hundred. Besides raising revenue, municipal operation of the herring fishery had an additional advantage since the fish were, at times, used to pay town debts. In 1683, the town compensated William Haskins for his services as Middleborough town clerk with "a load of fish taken to his house."

The rapid commodification of herring was reflected in the prices it fetched during the colonial era. From the outset, Middleborough fixed the market price of the fish in yet another means of establishing control over the local fishery. In 1698, three pence per thousand was set as the going rate. The following year, the fish were sold for eighteen pence per load, and the price continued to fluctuate thereafter between six and twelve pence per load. The growing demand for herring during the second decade of the eighteenth century, fueled by the arrival of new settlers and the expansion of existing families in Middleborough, was instrumental in subsequently driving up prices, which between 1709 and 1713 doubled to twelve pence per load. By 1716, the price of herring had risen to sixteen pence per load, and from there it maintained its steady inflationary climb, reaching eighteen pence in 1720, two shillings in 1721, three shillings in 1723 and six shillings in 1746.

Problems with the Colonial Herring Fishery

A number of problems plagued the operation of Middleborough's early herring fishery. Although town control was designed to sensibly exploit the resource while protecting it for future use, the actions of residents were not always in accord with this policy.

From the outset, the town strove to maintain an equitable distribution of fish. One of the charges to its herring agent in 1705, Joseph Barden, was "to take care that each man shall have his turn" at the weir when securing his load as allocated by the town. Nonetheless, misunderstandings arose in regard to what constituted "a load." Initially, residents had been entrusted to load their own wagons. By 1714, no doubt in response to the scene of excessively filled wagons brimming with herring drawing away from the river, the town felt compelled to appoint an agent who would assume responsibility for the loading.

Over the subsequent years, the town alternated between packing loads for residents (1720) and requiring residents to do the work themselves (1717), but the problem of residents taking more than their share persisted: "Heretofore a load of fish was any amount that a man's team could draw away. Some men being selfish began to make tremendous loads." Ultimately, in 1725, in an effort to standardize the procedure, a load was determined to contain eight thousand fish.

A second problem involved residents who were delinquent with their payments to the town. As early as 1702, residents were offered herring "on credit," being permitted to pay eight pence per load at the time of taking or twelve pence before July 31. The credit period was extended the following year to October 31. However, continuing problems were encountered by the town with late payments. In 1707, after some residents had been remiss in their payments for the previous year, a penalty fee of three pence was added to the charge for the fish if the sums were not paid within a month of taking the herring. Finally, in 1715, it was ordered that payment be made upon delivery. Yet on this issue as well the town seemed to vacillate between periods of strictness (demanding immediate payment at the time the fish were taken) and leniency (allowing residents to pay at some later date). In 1720, Middleborough required residents to pay for their fish at the time of removal and accepted either bills of credit or "three pecks of Indian corn, per load, when it was matured." Although the town permitted payment to be made "on trust" the following year, cash was again being required in 1722 and 1723 as payment up front prior to removal. The demand for cash may have been partially motivated by the town's need for funds to defray the expense of a new weir. In 1727, residents were once more permitted to take fish without payment, although the expense was added to their subsequent tax assessment.

Herring Agents, Speculators, Embezzlers and Poachers

To carry out its orders and maintain some semblance of regulation over the local fishery, the potential lucrativeness of which attracted Middleborough's less scrupulous characters, the town periodically appointed herring agents. The role of the agent was to "superintend the catching and the distribution of the fish, to collect the money due, and to see that the fish were properly guarded, and that none were caught except at the weir, by those authorized by the town, and at times appointed."

In 1696, the town selected three men to serve as agents, "to use all lawful means for the taking of the fish for the benefit of the whole town." Two years later, John Howland and David Thomas were named as agents to maintain the free passage of fish up the river and to oversee their careful taking as they came down the river, as well as that "they be equally divided among the inhabitants of the town."

Agents appointed throughout the era were tasked with various duties. In 1700, John Allyn and Samuel Pratt were named to fence the weir in order to prevent free-roaming swine from damaging it. David Wood and Samuel Pratt were named in 1702 to take the fish. Joseph Barden was named for a similar role in 1705. In 1722, the "Town agreed with Thomas Nelson that he should take his fish at the ware that now is at Assawampsett Brook & load the carts, for 12 pence a load."

At times, the colonial herring fishery brought out the worst in local residents, including the agents themselves, some of whom either sought to circumvent the various regulations governing operation of the fishery or ignored their responsibilities to the community. The agent during the late 1720s appears to have been negligent in forwarding the town the monies received on its behalf from residents taking herring at the weir. In 1731, the town voted to demand receipt of the monies taken over the previous four years from the agent, "and if he refuse, to prosecute for same." A later chronicler noted the incident as "an early instance of a defaulter and embezzlement of public funds." It was likely one of Middleborough's first instances of public fraud.

Financial speculation in herring remained a persistent issue, with entrepreneurially minded residents taking fish with the intention of selling them elsewhere for profit, a development attributable to steadily progressing herring prices throughout the period. In 1711, the town voted to prohibit sales by individuals of Nemasket herring outside town upon penalty of a fine of five pence per hundred sold. Nevertheless, black market sales continued contrary to the 1711 vote, and the order that no man be permitted to carry away fish to sell to others had to be renewed in 1722 in an effort to stave off growing speculation. Still later, in 1728, residents were prohibited from taking and barreling fish for sale on penalty of twenty shillings.

Speculative buying was also encouraged by the fact that by 1730, Middleborough herring had earned for themselves a widespread reputation

Herring River, Sandwich, Massachusetts, early 1900s. This was the first Massachusetts river to receive legal protection for its herring. The Nemasket was the second. *Author's collection.*

and were much in demand outside town. Almanac publisher Nathaniel Ames of Dedham, who had been raised in neighboring Bridgewater, wrote in several popular periodicals that he published between 1726 and 1764 of the Nemasket's most famous inhabitant, doing much to popularize Middleborough herring. In his notes for March 1730, Ames informed readers, "By-and-by Alewives (a sort of fish) will come, as much as I could wish. I think next week then I shall have them sure. Where is the place, you ask me? I answer, I think some folks call its name Namasket (river)."

Similarly, in March 1735, Ames wrote that "at this time of year Namasket River is a market place." No doubt some felt that such publicity was unwanted, encouraging as it did profiteering in herring that ultimately was conducted to the disadvantage of the town.

Poaching, too, was a serious problem, and despite efforts designed to protect the Nemasket fishery from this activity, the practice continued. In 1749, it was noted that "there are great quantities of the fish called alewives, which pass up the rivers and brooks in the town of Middleborough to cast their spawn; and not withstanding the penalties annexed to the many good and wholesome laws of this province already made to prevent the destruction of alewives, yet many ill-minded and disorderly persons are not deterred therefrom." In reponse to the situation of poaching at Middleborough, the Massachusetts legislature on December 23, 1749, passed a strict law calling for harsh penalties for those convicted of compromising the Nemasket fishery in any way. The Nemasket thus became only the second river in the commonwealth so protected—Sandwich's Herring River, troubled by poachers who hid in the thick woods that covered the river's banks and so escaped detection, had been protected in 1745–46.

Penalties for infractions were severe. Middleborough children and servants found in violation of the law were to be "punished by whipping, not exceeding 5 stripes, or by being put in the stocks, not exceeding 24 hours, or imprisonment, not exceeding 24 hours"; they themselves, their parents or their masters made restitution.

Colonial Industrialization and Herring

Because of its economic value to the community, herring required protection, not only from poaching and greed but also from industrialization that since the erection of the first "water mill" in New England on the Neponset

Star Mill dam, 1950s. By 1700, the causality between obstructed rivers and destroyed herring runs was well understood. Suitable passages permitting herring to ascend dams became critical. *Middleborough Public Library.*

River at Dorchester in 1634 had been rapidly transforming the face of New England. Industrialization would subsequently pit the individual rights of property owners whose dams obstructed the herring and potentially destroyed runs against the communal rights of towns like Middleborough, with inland fisheries that sought to preserve the herring for agricultural use. In an effort to mediate the growing differences, colonial legislation proved pivotal.

Some towns would be riven by the conflict between industry and fishery. In Taunton, where the community had voted that the "fish should not be stopped," provision initially had been made to idle the Cohannet mills during the spring months to accommodate the herring's spawning run. In 1664 and again in 1701, however, complaints were lodged that the then owners of the mills were failing to comply. In the latter year, much bitterness was created between those who favored ensuring passage for herring and those led by dam owner Robert Crossman, who favored industrialization. Crossman characterized fishery proponents as "those that lust after fish" and taunted them to stop their "fishy noise."

Perhaps the most notorious instance of factionalization was the so-called "herring war" at Falmouth, where three mills on the Coonamessett River had contributed to the decline of the local herring fishery. After a group of residents unsuccessfully sought to remedy the situation, indicating that "certain obstructions must be removed, some mill privileges interfered with, individual rights must be invaded, &c.," opponents of the fishery in 1806 attempted to demonstrate their contempt for the fish and its supporters by stuffing a cannon located on the village green full of herring and setting it alight. The cannon exploded, killing the gunner. Athough the tragic outcome temporarily defused the situation, the conflict between fishery proponents and industrialists remained unresolved, both at Falmouth and elsewhere.

The debate over the fate of the herring in these and other communities revealed a previously unthinkable assault on the traditional view of herring as critical to the well-being of the community. Nearly a century following the landing of the Pilgrims, some residents of southeastern Massachusetts were beginning to challenge this view, questioning the continuing relevance of the fish in light of the overall economic progress that had been made since 1620. At Middleborough, this challenge would be met differently.

Early Industrial Development along the Nemasket

The first dam on the Nemasket River was erected prior to 1670 upstream from the Wading Place in the vicinity of the original native fishing weir and was followed by dams at Muttock (circa 1676), Wareham Street (1762) and Murdock Street (circa 1802). Little is known of the original Nemasket dam. It was said to have been an earthen dam about 250 feet in length and 5 feet high and situated some 300 feet upstream of the later 1833 Star Mill dam. Reputedly, the dam had a very shallow impoundment, barely capable of powering the gristmill that operated there. It has been recorded that the Middleborough dam did not hinder the progress of the fish either up or downstream, so a fishway was likely provided in the form of a natural, slowly inclined stream similar to what would exist near the site some three centuries later.

The dam was probably neglected for a period following King Philip's War, but with its reconstruction in about 1679, the town became concerned about maintaining the local herring run. On June 18, 1686, the town named

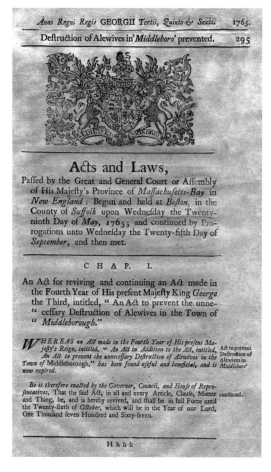

Act to Prevent the Unnecessary Destruction of Alewives in the Town of Middleborough, 1765. Middleborough successfully upheld herring laws like this. *Author's collection.*

a committee "to treat with Capt. John Williams about his *mill* & also for the fish that they be preserved, that both may be ordered as may be both for the good of the Town & beneficial to the owner." Here the town recognized not only the importance of promoting industrialization but of protecting its herring fishery as well.

Following 1686, the effort to balance the needs of industry with the desire to protect the fishery would be a constant challenge for the town, yet the conservatism of the town and its emphasis on agriculture allowed Middleborough to avoid the contentiousness witnessed elsewhere. Emboldened by the absence of a strong pro-mill party, the town in March 1695 joined residents of Plymouth in petitioning the Massachusetts legislature to prohibit the construction of dams and mills that would interrupt the annual migration of herring, "they being beneficial to agriculture."

Shortly afterward, William Briggs Jr. in nearby Taunton wrote admiringly of Middleborough's reluctance to permit the damming of the Nemasket in contrast to Taunton, where dams along the Mill River would ultimately destroy its fishery:

> *It is well known how much other Towns are advantaged by this sort of fish. Middleboro will not permit any dam for any sort of mills to be made across*

their river to stop the course of fish nor would they part with the privilege
of the fish if any would give them a thousand pounds and wonder at ye
neighboring town of Taunton, that suffer themselves to be deprived of so
great a privilege.

In 1734, when proposals called for the raising and strengthening of the Muttock dam in order to support enlarged industrial operations there, "objections were made…because it would interfere with the catching of herring…and so destroy multitudes of the young fish." Despite these objections, a rolling and slitting mill was constructed, but accommodation was made for the herring by deliberately idling the works each spring. In 1743, specific mention is made of the necessity of stopping the works for two months in order to let the fish upstream, and the dam's subsequent owner, Peter Oliver, later noted that he also had to cease operations during the spring in order to leave the dam open for the fish. Similarly, the ironworks at Wareham Street operated seasonally, work being suspended during the annual herring run in order to permit passage of the fish.

Early owners of these works were compelled to provide passage by a series of new laws regulating herring fisheries. The 1735 law passed by the Massachusetts legislature, concurrent with the improvement of the Muttock dam, stipulated that any dam constructed on a herring stream had to be provided with a fishway, and that fishway had to be open during the period of the upstream migration. It was the first law passed in New England that afforded protection specifically for herring. Unlike elsewhere, however, where laxity in enforcing these herring-related laws would be cited as a reason for the decline of the fish, Middleborough achieved success in upholding colonial herring laws and, as a result, managed its fishery in a relatively sustainable way through the nineteenth century, when obstructions, overfishing and pollution began negatively affecting fish populations.

THE ECONOMICS OF THE
NEMASKET FISHERY, 1775–1900

COMMERCIAL FISHERY AND AUCTIONS

Throughout the nineteenth century, the Nemasket herring fishery continued to be regarded as a valuable communal asset, but the manner in which it was exploited changed radically. While the Town of Middleborough initially operated the fishery itself, by the mid-eighteenth century it believed that the run could be operated more profitably by auctioning the privilege (or right to take fish from the river) to a single person, who would be free to dispose of the fish as he saw fit. While it has been written that Middleborough did not establish such a commercial fishery until 1792, the town had "voted to sell the privilege of catching the fish at auction to the highest bidder" as early as October 8, 1764. Nelson Finney was the high bidder that year, paying the town eighty pounds for the privilege. The annual herring auction subsequently became a rite of town life for the next two hundred years.

A number of the earliest herring auctions were noted for the presence of alcohol, surprisingly furnished by the town at taxpayer expense as a means of lowering financial restraint on the part of bidders. A record of October 6, 1789, notes the naming of a "vendue master" (auctioneer) and a vote that free liquor should be provided "to encourage the sale." Other votes record the earmarking of ten shillings for the purchase of liquor and compensating a local resident for the use of his home as both an auction locale and site for storing the drink. In fact, much greater amounts were expended by the town, providing alcohol with which to lubricate its auctions. In 1779, Hercules Weston was paid two pounds, ten shillings for "Liquor at the Sale of the Alewives," while Thomas Sproat was paid nearly twice as much the following year. Relative to these sums, the twelve pounds paid to Sproat for providing liquor at the 1781 auction was astonishing.

Nonetheless, too much alcohol could prove problematic; yet another record "reveals that with the free liquor at a herring sale, the crowd became too noisy to continue the sale, which had to be adjourned." Undoubtedly, riotous auctions fueled efforts to restrict the general sale and consumption of alcohol by advocates of temperance, which became a strong social and political movement locally in subsequent years.

Another issue arose in the 1790s when the town's authority to auction the privilege was challenged by some residents: "Doubts have arisen, whether the inhabitants of...Middleborough are authorized by law to agree with and hire any person or persons to take [herring], and sell them at the price stipulated by the law, and to account with the said inhabitants for the net proceeds." Ultimately, in early 1801, a legislative act sanctioning the practice was approved.

The town's object in auctioning the privilege was to secure as much for the fishery as possible, and this task fell to the herring agent. When Albert Thomas was elected to the position by the annual town meeting in March 1859, he was encouraged "to sell the herrings for what he can get." While this was the common practice, on some occasions, such as in 1883, the town believed that more could be obtained through *not* leasing the fishery and operataing it itself, reverting to the eighteenth-century practice of hiring men to do the work and conduct sales of freshly caught fish. Town operation of the fishery, however, remained the exception and was undertaken only when efforts to otherwise lease the fishery for an acceptable sum had been frustrated. Nonetheless, it long remained an alternative to commercial leasing and was considered as late as the 1950s, when the fishery went unleased for other reasons.

Those who successfully bid for the privilege signed a bond ensuring that they would pay the town, although payments frequently were made months (and in some instances years) following the auction, particularly in the depressed economic climate following the Revolution. This circumstance, although it allowed successful bidders to sell the fish prior to satisfying their obligation to the town, encouraged the speculative nature of the annual auction.

Because the Nemasket tended to outsize other area runs, Middleborough was generally able to get more for its herring privilege than communities elsewhere with smaller runs. Amounts received, however, fluctuated dramatically from year to year depending on prospects for that year's run. For lessees, the acquisition remained a gamble. If the run proved poor, they stood to lose considerably.

FISH ACCOUNT

Monies realized from the lease of the herring privileges were deposited in the Middleborough town treasury, which for years included a so-called fish or herring account.

Funds were most frequently received from the sale of rights to the fishery, although there were other sources as well. Nominal fees paid by residents to take herring, monies generated from the sale of the fish in years when the privilege was not auctioned, reimbursements from the several communities downstream on the Taunton River following 1863, rare town meeting appropriations and equally infrequent legal settlements from violators of the various laws designed to safeguard the fishery also helped fill the treasury fish account.

Expenditures covered a wider range of items. Wages for both the herring agents and the fishermen engaged to harvest alewives were the largest expense. In 1801, inspectors received $1.50 for every twenty-four hours of service. In 1889, Everett T. Lincoln, Joseph T. Wood and L.M. Alden received a combined $39.00 for their services as fish wardens. Additionally, the account covered the expenses of hiring an auctioneer to conduct the annual sale of the privileges.

The expense of supplies varied from year to year. In 1868, $11.00 was expended "for baskets, nets, and twine" and $0.90 for "boxes and lines," while in 1872, $4.02 was spent on twine, $9.50 spent on nets and $4.50 paid to David A. Tucker for tubs. W.H. Vaughan & Son received $4.75 in 1872 and $2.55 in 1885 for mending nets.

Legal services, too, could take a substantial bite from the herring account. Middleborough attorney Everett Robinson was paid $37.50 in 1871 for "services before Fish Committee" and $12.25 "for service before committee on fisheries in 1875"; W.H. Osborne was paid $17.90 for "legal services

Herring fishermen at the Star Mill, 1950s. For generations, these men were a presence along the Nemasket River. *Middleborough Public Library.*

before the Committee of the Legislature" in 1875; and Attorney John C. Sullivan received $10.00 "for trying fish case at Taunton" in 1878.

Although in some years large sums were realized from the sale of the fishery, often these receipts were offset by unanticipated expenses, such as the sixteen dollars expended in 1874 "in protecting young herrings." These unbudgeted costs could be enormous, as in 1883, when the construction of a new herring house at the Star Mill, the requirement of keeping a perpetual watch on that structure throughout the season, the increased expense for labor during the harvest, the need for additional salt for preserving the larger catch and investment in new nets, tubs and frames substantially eroded the town's profits.

The Nineteenth-Century Herring Market

As was the case in the past, the price of herring in the early 1800s continued to be set by the town, its authority to do so being confirmed by the

Massachusetts legislature in 1797. The price of alewives was fixed at 15 2/3 cents per 100 in 1796 (with 120 fish being counted as 100) but rose due to the influence of market forces. The local herring price was pegged at 25 cents for each 100 in 1797 and 1798, 75 cents per 100 in 1816, 50 cents in 1820, 42 cents in 1822 and 50 cents in 1845. The prices throughout the era were comparable with those elsewhere, where during the 1830s the fish were selling for $1.50 to $2.00 per barrel.

The yearly fluctuations in herring prices were attributable to a number of factors, including the size of the run, size of the fish, demand for the fish, availability of markets in the West Indies and economic factors such as the poor economic climate following 1813, which undoubtedly explains the high price quoted for 1816.

In May 1854, the fish were selling directly from the weirs at a price once more of 75 cents per 100. The following year, the early April catch was termed "fine, and well worth 16 cents per dozen, the price at which they are now sold." The higher price over the previous year was attributed to the fact that the herring taken in 1855 were believed to have been of a larger size. By May 1855, however, the price had fallen to 50 cents per 100. This was the price that was set at the annual town meetings in April 1856, March 1857 and March 1858. The latter meeting, however, voted that "persons may have their first 200 at 25 cents a hundred," possibly in view of the tight economic climate of the time. In 1860, the town once again voted to sell at a price not exceeding 50 cents per 100 and moved to ensure that herring were available for all, approving the restriction that "[n]o one to have over 200 when others are waiting, &c."

By 1861, however, circumstances had changed. With civil war imminent, the town meeting of that year (as well as in 1862) voted to sell herring for the highest price attainable, thereby submitting fully to the play of the market. In consequence, the price rose swiftly. By 1867, herring were selling at the weirs for $1.51 per 100. The postwar price peaked at $1.63 1/3 per 100 in 1868 and gradually fell from there to $1.45 4/10 per 100 (1869), $1.31 per 100 (1870) and $1.00 per 100 (1872). By 1875, the price had fallen to prewar levels for the first time since the war, and herring were "selling at Muttock for the low price of 75 cents per hundred," prompting the *Middleboro Gazette* to urge its readers, "now is the time for you to procure your quota."

Although the town continued to dictate the price for which herring were sold, in so doing it was forced to account for the influence of the market. In 1876, the late arrival of the herring forced the town to set lower prices than it had anticipated: "It being so late before the fish got here, and the market being supplied from other places, where they are taken much earlier, the price of the fish had to be made quite low." In the following year, 1877, the bottom completely dropped out of the market, and herring were being purchased by Middleboro farmers for 10 cents per 100.

Herring Houses

Until 1833, three separate Nemasket herring privileges were auctioned: near the industrial works at Muttock, at the stone weir in the vicinity of East Main Street and at Assawompsett Brook. Although a decision was taken in 1833 to lease a single privilege only, which the lessee was free to choose, Muttock had become the principal site for harvesting herring by mid-century.

Originally, fish were hauled from the river in great hand nets. At a time when Massachusetts restricted herring harvesting to the four days between Monday and Thursday, Mondays were considered the best days for taking fish, as none had been taken on the previous three days and the run consequently proved heavier. During the time when the fish were hand-netted, all but herring and shad were removed by hand and returned to the millpond. Shad were barreled separately and sold at a much higher price than the herring. "All of a sudden you might hear the hoarse cry 'Shad Ho!' and then you knew that not only did they have a netful of fish but that shad—sometimes called big herring—were caught up along with the rest." Inevitably removed during this process were lampreys, trout, bass, yellow perch, eels, pickerel and small turtles.

In the years that the town retained the privilege and harvested the fish itself, it engaged men to do the work. In 1867, Joseph H. Bisbee and Ezra A. Harlow were hired for this purpose, being paid $99.59 and $73.25, respectively. Bisbee, as owner of the Muttock gristmill since 1861, was perhaps a convenient choice, and he undoubtedly took the fish there, while Harlow was engaged at the Lower Factory (East Main Street). James H. Waterman was commonly employed by the town to harvest herring as well. The fish caught in these years were sold directly by the town at the weirs, with fish dealers or middlemen making their purchases there.

The taking of herring by the town required facilities to process and preserve the fish, either through smoking or salting. In time, herring houses were erected near the two principal fishing pools at Muttock and the Lower Factory. Here

Muttock herring house (center), circa 1900. *Author's collection.*

fish were smoked and salted and kept for distribution to residents eligible for free fish. According to Muttock resident James A. Burgess, his father "every year cured and dried 30,000 herring for the people of the town."

Drying herring involved first cleaning the fish by removing the scales and viscera and then pickling the eviscerated fish in brine—tasks often done in a separate building known as a fish scale house. Following brining, the fish would be rinsed in cold water and left to dry in a spot out of the sun, preferably one with a breeze. Then the fish would be hung on sticks that were passed through the fishes' eye sockets. Frequently, children would be engaged in the task of placing a dozen fish on a stick preparatory to smoking, earning a penny for each completed stick. The sticks would then be suspended in the herring house for curing.

The length of curing varied, although generally five days was the rule for those fish intended to be preserved for a long period of time. Fish were smoked until they turned an even bronze color. In 1807, Maria Eliza Rundell, in her *A New System of Domestic Cookery*, an early American cookbook, outlined the method for smoking herring:

Clean, and lay them in salt and a little saltpeter one night, then hang them on a stick, through the eyes, in a row. Have ready an old cask, on which put some sawdust, and in the midst of it a heater red-hot; fix the stick over the smoke, and let them remain 24 hours.

Alternatively, the fish could be salted, a process that consumed enormous quantities of salt. During 1857, the large fishery on Martha's Vineyard utilized so much of the article processing herring for the southern market that "about all the salt on the Vineyard is used up." In 1883, the Town of Middleborough paid grocer Matthew H. Cushing the then large sum of $44.10 for salt with which to preserve its catch that year (more than 3 cents per 100 fish).

The Muttock herring house where herring were preserved was located on the right bank of the river, where the parking lot for Oliver Mill Park is now located, and the three-dollar rent for the land on which the house stood was was paid annually to the Sproat family. There is little history of the final demise of the Muttock herring house. It was stated to have been destroyed by fire, and the fact that no rental payments were made by the town to the Sproat heirs following 1901 indicates that the building was likely destroyed about this time, a period during which the entire Muttock site was falling into general disuse as a fishing pool.

The town also operated a second herring house at the Lower Factory. Built in 1833, this herring house fell into gradual disrepair, and by 1877, the need for a new structure was recognized, as "people will not subject themselves to the inconvenience of going and waiting at the weirs as formerly, when they can have the fish brought to their doors, all cured, for $1.00 or less per 100."

In 1882, Middleborough voted to erect a new herring house at the Lower Factory, but its location there proved contentious since the Star Mill, which then owned the site, disputed the town's right to locate it on the company's property. "There is a dam row about herrings at Middleboro, the Star Mills insisting that what the town claims as a right is only a privilege," summarized the Plymouth *Old Colony Memorial*. Confident that it had the upper hand, the Star Mill demolished the town's herring house in early March 1884 shortly before the run was to begin. "The fish-house, so long used by the town at the Star Mills, has been torn down by the company." The *Middleboro News* noted that the unilateral action on the part of the woolen manufacturer "may mean a lawsuit." While such appears not to have been the case, it did prompt the town to act. Despite the company's objections, a herring house was quickly rebuilt alongside the river at an expense of $39.84, probably sometime in March. An additional $105.60, however, had to be expended in

Muttock herring house (far right), late nineteenth century. *Author's collection.*

order to mount a guard "on account of the Star Mills authorities forbidding a building of any kind to be placed on any part of their premises, and the positive assurance, both before and after the building was built, that any such building would be torn down." E.A. Harlow, Benjamin W. Bump and A. Baxter kept a continual watch on the herring house during the 1884 run, at the conclusion of which the town "took the building away" so as not to further antagonize the Star Mill.

MAINTAINING THE NEMASKET FISHWAYS

Increasingly throughout the nineteenth century, industrial works along the Nemasket River grew in economic importance. By 1854, Middleborough was producing 415,000 yards of printing cotton (valued at $16,000) at two factories located along the river, as well as $49,500 worth of shovels, saddles, harnesses and trunks, a good proportion of which was attributable to the manufactories at the Lower Factory and Muttock. In contrast, the herring fishery realized only $1,800. Yet while industry clearly offered a greater financial return, residents believed that if managed properly, industry and fishery together could provide maximum economic benefit to the town.

At Muttock—where a forge, an iron and coal house, a slitting mill, a bolting mill, a trip hammer shop, a sawmill, a gristmill and a shovel works operated following 1797 under the ownership of Abiel Washburn and Thomas Weston—provision was made for the annual passage of alewives, possibly in the form of a fishladder. Tellingly, an entry in the Middleborough treasurer's record book for the year 1798 lists the Muttock site as including what was described as the slitting mill "ditch," indicating perhaps that a fishway had been provided in the form of a crude inclined cut. Alternatively, the fishway may have been simply an "old canal…ostensibly employed as an outlet for the overflow of the mill pond," as suggested by the *New York Times* in 1897.

Throughout the period, despite the frequently high level of industrial activity at the site, Muttock remained the principal location at which herring were taken in Middleborough. Combined with the busy shovel manufacturing and milling operations, the annual spawning run with its attendant wagons, nets and fishermen often produced scenes of congestion, as noted by James Burgess:

> *I have seen of a Monday morning more than 40 teams at 4 o'clock waiting for their turn to get some herring…. I have seen of a Monday morning 40,000 herring caught at Muttock dam before 12 o'clock.*

The abandonment of Muttock by the Washburns following 1850 left the Sproat family the most active players at the site, but they had neither the inclination nor the financial wherewithal to arrest the economic and physical decline into which the industrial works, including the dam, were slipping. Following severe rain and heavy winds, the "herring weir dam" at Muttock was swept away during a freshet on February 9, 1867. A subsequent structure was washed away in another freshet on St. Valentine's Day 1886, and its replacement, too, seems to have broken down. In 1890, Randall Hathaway was paid $23.25 for "lumber and labor for fixing dam at Muttock," and by the first years of the following century, Muttock residents were lobbying for the construction of a replacement.

During the last quarter of the nineteenth century, the Muttock site, abandoned by industry, rapidly became overgrown with brush, prompting some lessees of the annual herring privilege to harvest at the Star Mill.

Muttock, 1890s. Following the destruction of the Muttock gristmill by arson in November 1886, the site was abandoned and left to decay. *Author's collection.*

Nonetheless, an 1897 *New York Times* article noted that herring were still being taken at Muttock: "Near a ruined mill nets and dams are carefully arranged in a canal which rounds the falls to trap the little travelers on their way up the river…. Here the men hired by the contractor to capture the fish camp out in a shanty…The biggest catch is made at Muttock, the maximum being 100 barrels a day." Shortly after this article appeared, however, the Muttock privilege seems to have been abandoned completely. In 1911, when the Gloucester Fish Company made use of the Muttock privilege, it was noted as "the first season for a long time."

Industrial development at the next dam upstream from Muttock initially appeared incompatible with maintaining the rights of the herring when, in

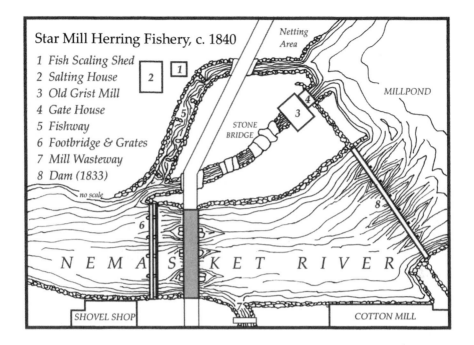

Star Mill Herring Fishery, c. 1840

1 Fish Scaling Shed
2 Salting House
3 Old Grist Mill
4 Gate House
5 Fishway
6 Footbridge & Grates
7 Mill Wasteway
8 Dam (1833)

no scale

Netting Area

MILLPOND

STONE BRIDGE

N E M A S K E T   R I V E R

SHOVEL SHOP

COTTON MILL

1812, a cotton manufactory was built on the west bank of the river near East Main Street at what came to be known as the Lower Factory. It is likely that the cotton mill operations ceased in the spring, as was customary to permit passage of the fish, although there appears to be no record of this at the site.

In 1833, when Horatio G. Wood and Peter H. Peirce of Middleborough proposed relocating the Lower Factory dam farther downstream to power a new cotton mill, the town voted not to oppose the relocation, provided Peirce & Wood "make and always maintain a sufficient passage-way for alewives to pass up around the said dam, at the suitable season of the year." Provision for the herring was made in the form of a natural fishway on the south end of the dam. An entrance into the ladder was created just above the dam, with the ladder descending on the southern side of the river toward an outlet immediately below a small private bridge that crossed the Nemasket.

Despite the later reluctance of Peirce & Wood's successor, the Star Mill, to permit herring fishing on its property, the company did not oppose late nineteenth-century municipally funded improvements to the fishladder, presumably because such improvements reduced the number of the fish passing through the company's turbines. In 1878, the town meeting considered whether "to repair and provide better accommodations, at the fishway, for catching the fish at the Star Mills," while in 1880, Tucker & Dunham were paid $21.70 "for repairing fish-way at Nemasket." Despite

Nemasket River at Murdock Street, early 1900s. *Author's collection.*

these alterations, the Star Mill fishway in 1897 remained "a natural brook, with many pretty cataracts and miniature waterfalls, through and over which the herring sport in high glee."

The third of the four nineteenth-century dams located along the Nemasket River was situated at Warren's Mills, a site located immediately downstream from Murdock Street and developed sometime between 1802 and 1812, when a dam was erected by Beza Tucker of Boston and Daniel Warren of Middleborough. By 1822, a gristmill, a sawmill and a forge that included a "stove and shed" were operating at the site. In 1867, the mills, by then known as Warren's Mills, were purchased by Zebedee Leonard, who in 1870 petitioned for leave to erect a fishway around the dam, signaling an intention either to raise the dam or to mill year-round. It is not clear whether Leonard's petition was approved, nor is it known whether a fishway ever existed at Warren's Mills. What is known is that accommodation for the alewives was provided as required by law either by means of an undocumented fishway or by deliberately idling the mills each spring. After 1880, when an incendiary fire in autumn 1880 destroyed the sawmill, all industrial operations at Warren's Mills ceased. In time, either decay or removal of the Warren's Mills dam permitted the alewives free passage at Murdock Street.

UPPER FACTORY PRIVILEGE

In contrast to industrial sites elsewhere along the Nemasket, the Upper Factory at Wareham Street was a source of continual frustration relative to the herring. So plagued was the town by the issue of the Wareham Street fishway that a local newspaper in 1905 characterized the question as one "which like Banquo's ghost will not down." The persistence of this conflict was partly attributable to the fact that it was at Wareham Street dam that fishery and industrial interests clashed most markedly, mirroring conflict in other communities.

Economic opportunities and technological advances at the start of the nineteenth century encouraged construction of a textile manufactory in 1813 at the Upper Factory by the New Market Manufacturing Company, which was formed to manufacture cotton and woolen cloth, yarn and iron goods. By mid-century, a box mill, a gristmill and a sawmill were also operating at the site.

It was not until 1867, however, that fishery interests were threatened when proposed expansion of Brown, Sherman & Washburn's shovel and hammer shops threatened to impinge on the existing fishway with the potential of disturbing the herring. "It is believed that the herrings will meet with fewer obstructions if a new canal is cut thro' a few rods east of the works," helpfully suggested the *Middleboro Gazette and Old Colony Advertiser* in response.

Although this proposal was adopted, work on the ladder was delayed by the destruction of Brown & Sherman in a November 1868 fire. The ladder may have not been constructed immediately in 1869 and possibly dated from as late as 1873, when the *Middleboro Gazette* described it as "new." The ladder was reported in the newspaper at that time as having been "built upon scientific principles, and it seems to be appreciated by the herrings, if we can judge by the thousands that were wending their way up the circuitous route…. They glided along with comparative ease and winked their eyes at us in a way that intimated that they were well satisfied with the way things were fixed and wished to offer a vote of thanks to Mr. Sherman."

According to a physical description printed in the Plymouth *Old Colony Memorial*, the new ladder was "a trough about four feet wide leading from the bottom of the river, below the dam to the top. At intervals of about five feet, planks, for rests, are placed crosswise through which, at one end, holes are made for the fish to pass through." The *Middleboro Gazette* remained hopeful regarding the new arrangement, stating that "it seems to be what is needed, and we do not think that after this, there will be any more trouble at that

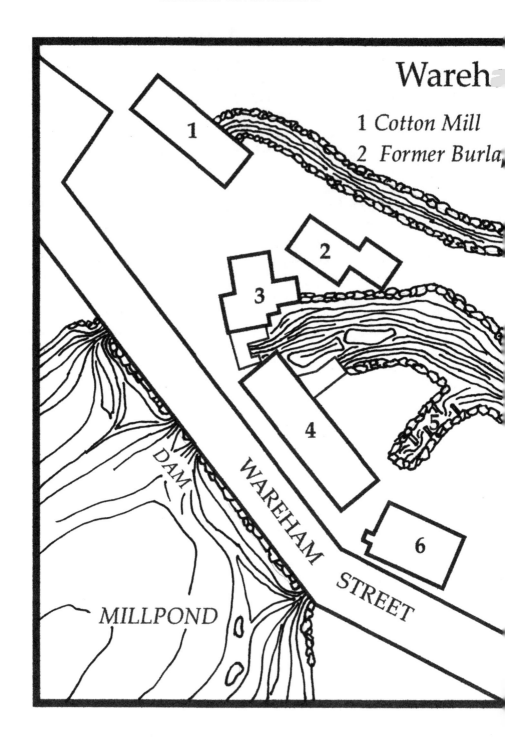

Wareh

1 Cotton Mill
2 Former Burla

treet (Upper Factory) c. 1875

3 Sawmill

4 Shovel Works

5 Fishway

6 Box Mill

7 Blacksmith Shop

8 Sampson House

9 Shovel Shops

10 Cotton Mill Wasteway

NEMASKET RIVER

place about the passage way for herrings, as they appear now to get up with comparative ease."

Little maintenance appears to have been provided for the fishladder following its construction, and consequently both it and the dam were in constant need of repair. The site as a whole was increasingly neglected, particularly following June 1875, when the cotton mill owned by Sherman and occupied by A.B. and J.A. Sanford was destroyed by arson. By 1878, complaints were being lodged regarding the condition of the ladder that may, in some part, have been attributable to deterioration of the overall dam structure itself. In the spring of that year, some thirty thousand herring had to be put over the dam by hand due to the inadequacy of the fishladder. Although the selectmen optimistically wrote in the first months of 1879 that "we are also assured that the fish way at the shovel works shall be attended to directly," it would, in fact, be another four years before the fishladder was repaired.

The situation at the ladder in 1878 came to the attention of State Fish commissioner Edward A. Brackett, who "pronounced the one at the upper works as being decidedly wrong." As a result of Brackett's assessment, the town meeting of 1880 voted to construct a new fishway on the southeast end of the dam "beyond all of the works," similar to the one at the Lower Factory, proposing a three-hundred-foot-long, six-foot-wide ladder with a drop of thirty feet. Disagreement regarding funding and maintenance, however, delayed the project, and in the spring of 1883, the ladder was described as impassable, requiring the transfer of some 5,300 alewives over the dam.

Finally, in 1883, the defective ladder was replaced with a new style of fishway developed by Brackett known as the Brackett fishway. But not all were satisfied with the result. Longtime Middleborough fish warden James A. Burgess held strong opinions about how herring should be protected and was dismissive of those who held differing views, including Brackett. "When Sherman and Brown bought the upper dam our troubles began [and] they exist still," Burgess wrote in 1908. Burgess contended that Sherman, in his drive to expand the works, had sought to control as much of the flow coming over the dam as possible and had contacted Brackett to view the site and draft a solution to satisfy the needs of the shovel works, resulting in the installation of the Brackett fishway.

In Sherman's defense, he may have been inclined to cooperate with Brackett, who in 1864 convinced the Charlestown Water commissioners to remove their dam on the Aberjona River, much to the detriment of the Fibrilla Flax Mills, a linen manufacturer that drew power from the river and had frequently closed the dam's fishway when it required additional water

Brackett fish ladder, East Taunton, circa 1900. The nearly identical Brackett ladder at Wareham Street was criticized by some as a "stupendous failure." *Author's collection.*

Former shovel works site at Wareham Street, late nineteenth century. *Author's collection.*

power. On May 7 of that year, representatives of the water commissioners, "equipped with a couple of kegs of gunpowder," blew the dam to bits in complete disregard of the manufacturer. Ironically, the removal of the dam ultimately proved immaterial as far as herring were concerned; by the 1890s, pollution in the Aberjona and Mystic Rivers had killed all the fish.

Possibly aware of Brackett's history on the Aberjona, Burgess was thoroughly unimpressed with both the man and his ladder. "Mr. Brackett, the chairman, came out and he had, he said, invented a patent whirligig which if he was allowed to put it in would beat nature itself.... And we can go today and look into that old Brackett fishway and see what a stupendous failure it was." (When Brackett retired from fisheries work to crossbreed poultry on his Winchester farm, the sharp-tongued Burgess mocked him for "sitting Rhode Island red bantams on Mongolian pheasants' eggs.")

Although the Wareham Street fishway did ultimately prove problematic, initially issues at the site stemmed more from a failure to keep the fishway in good repair rather than any immediate deficiency with its design. The condition of the fishway was symptomatic of the overall decay into which the Upper Factory site was slipping. In January 1884, Sherman's shovel works building burned, following which the remaining smaller buildings were vacated. Soon afterward, they were characterized as "old and dilapidated."

Another problem at Wareham Street was the issue of the fishway not being located by some herring that, as a result, became trapped in the river below the dams. This development arose from the herring's natural instinct to follow the greatest flow of water (which at the time of the spawning run generally poured over the tops of the dams). As early as 1876, the town recognized that "this year that at least 15,000 fish have gone up through the fishway at the Star Mill, and not finding the fishway at the Shovel Works, have gone back, and many were taken in the nets going down the fishway, while many more have gone over the dam and been confined between the dam and the grates." In both 1890 and 1891, the town went to the considerable trouble of "putting herring above the Shovel Works dam" by netting them below the dam and releasing them into the millpond above. Finally, in 1892, mud sills and grating were installed.

Still yet another issue with the passage of the fish had to do with mature herring returning to the sea after spawning, followed by their fry a few months later. Particularly concerning were the turbines of the Star Mill and the water wheels of the shovel works at Wareham Street. It was remarked in 1876 that "considerable complaint has been made in years past, that large quantities of young fish were killed in passing down through the dams, many of them obliged to pass through the iron wheels, and especially so the past year, on account of scarcity of water," which forced the fish to follow the sole channel through the factory intake or wheel raceways.

Manufacturing Discouraged

It is not clear to just what extent the community's desire to protect its fishery may have deterred interested parties from locating industrial works along the Nemasket, although there was certainly a negative impact. During the eighteenth century, when not all of the Nemasket dams were outfitted with fishways, the requirement to temporarily suspend industrial operations for two months each year may have been regarded as onerous by some, particularly since the prodigious growth of grass in the river could force a second suspension of operations at the height of summer. Yet even the powerful Peter Oliver, who appears to have been displeased about having to stop his works at Muttock each March and April, was unwilling to challenge the town's established priorities regarding herring.

The Nemasket was not the only river in the region where herring were prioritized over industry. The undammed Mattapoisett was another. "The fishery is owned by the towns of Mattapoisett, Marion and Rochester, and it is stated the [industrial] privilege could be sold for $40,000, could consent be obtained to destroy the run of fish. This the towns wisely refuse," noted the *Old Colony Memorial* in April 1876.

Weymouth, like Middleborough, traditionally had been protective of its herring fishery, authorizing a committee in 1724–25 to treat with mill owners on the Weymouth Back River with a view of providing a passage for the fish upriver. In 1822 and again in 1824, voters in Weymouth refused to lease the privilege, a decision marked according to one later commentator by "strange short-sightedness." Although the following year the town agreed to permit mill development along the river, provided a sufficient fishway be constructed to allow passage of the herring, "difficulties were thrown in the way and it was never carried out."

Pembroke historian Henry Wheatland Litchfield in 1909 reported a Pembroke tradition that North Easton shovel manufacturer Oliver Ames had been refused a privilege on Herring Brook for fear that shovel manufacturing would jeopardize the herring, with an investigating committee reporting in 1838 that "the destruction of the herring is

Mattapoisett River privilege, circa 1900. The Mattapoisett herring survived largely due to a reluctance to permit industry along the river's course. *Author's collection.*

Pembroke herring run, circa 1910. Like Middleborough, Pembroke prioritized its herring over industry and maintained a viable run. *Author's collection.*

inadvisable." Writing in hindsight nearly seventy-five years after this refusal, Litchfield was able to perceive advantages prioritization of the fishery over industry had had: "We may even think [Pembroke's] vitality suffered from her preference of herring to shovels. But surely something is due the now despised alewife, that preserved for us an autumn landscape of purple hills and russet meadow unequalled in all the country round."

While manufacturers were not denied outright industrial privileges along the Nemasket, restrictions on their dams may have discouraged their locating in town. Although fishladders allowed manufacturers to avoid intentionally idling their works each spring, Brown & Sherman's experience at Wareham Street proved that ladders were not always an effective solution and that the rights of the fish could still conflict with the interests of industry.

Most attractive to manufacturers of the Nemasket privileges was the disused privilege at Muttock that was investigated a number of times as a potential industrial site following the Civil War. In July 1874, it was reported that an offer of $15,000 had been made for the Muttock privilege "by parties desiring it for manufacturing purposes." Eight years later, in December 1882, the prospect of a paper and pulp mill locating at the site was reported. Work proceeded so far as to include an analysis of the Nemasket River's waters, which were "found to contain nothing injurious to the product of such a mill." In late February 1883, the possibility of a thread mill being located at Muttock just downstream from the herring house was also being reported locally. This last rumor circulated for a month, and Middleborough residents were led to believe that the Willimantic Thread Company of Connecticut "will erect mills there at once." In 1923, sparked by a proposal by MIT students, the town investigated the possibility of constructing a hydroelectric plant at Muttock. None of these proposals came to fruition, and as a consequence, Muttock was left to revert back to its natural state.

While continuing protection of the herring may have convinced manufacturers to locate elsewhere, the application of steam technology to industry freed them from reliance on water power to drive their wheels. For the first time, industries were no longer required to locate along rivers, and the postbellum period began to witness abandonment of industrial sites along the Nemasket. While this development removed one threat to the herring migration, continuing use of the river as a conduit for carrying waste away from both existing manufactories and new municipal sources would continue to pose an industrial threat to the Nemasket alewife, a situation that would reach crisis proportions during the following century.

THE POLITICS OF THE NEMASKET FISHERY, 1801–1900

Tax Relief

The economic importance of the Nemasket herring fishery imbued the fish with a corresponding political importance, though one that changed over time. Once regarded as a vital factor in the success of English agriculture, by the nineteenth century, herring were increasingly valued for the funds they brought into the town treasury and for the ends to which these funds were directed: tax relief and social welfare.

Regarding the herring's role in reducing taxes, the local *Namasket Gazette* rhapsodized, "They are a delicious morsel naturally, but taste a little sweeter on reflecting that every mouthful eaten goes so far to pay our tax." In 1854, earmarking herring receipts for tax relief resulted in a savings of ninety cents for each taxpayer, then a substantial sum. Eventually, by 1904, the practice of using these proceeds to reduce residents' tax liability was abandoned, partly in response to a decline in the local fishery.

While receipts from the sale of herring were devoted to tax relief, the annual spring town meeting that disposed of the herring privilege remained of much interest to tax payers such as James A. Thomas, who in March 1909 admitted to looking forward to these warrant articles "with much pleasure." In 1922, it was stated that "one does not have to be very old to remember the dramatic moment in the annual town meeting when the herring privilege was auctioned to the highest bidder."

The annual auction of the herring privilege was often marked by much discussion, as well as frequent levity. During a special town meeting held in October 1861, "spirited remarks having herrings for a text, were made by several voters," while "the herring question, as usual, created some very lively talk, and some sharp jokes were cracked," noted an observer.

Another observer at Middleborough's 1860 herring auction remarked, "The herrings usually furnish plenty of '*bones* of contention' at our town meetings, but the question was settled this year with an unusually small amount of *buncombe*." Of bunkum there was certainly plenty at every town meeting, but following the mid-nineteenth century, Middleborough grew extremely consistent in its handling of the herring fishery. Indicative of this mounting trend was the comment made following the town meeting of 1862 that "the herring question was settled with an unusually small amount of talk." The growing consensus among town meeting voters regarding the management of the town fishery may have, in fact, signaled its decreasing relevance to taxpayers. As the receipts from the sale of the privilege plummeted particularly following 1898, residents seem to have simply lost interest in the face of more pressing issues, including fire protection, education and a municipal water supply. Nonetheless, the herring auction still had the power to rouse bleary-eyed voters from dryer, more complex matters. In 1905, "'Herrings! Herrings!' was the cry from various parts of the hall, at this point, and at the eleventh hour the fishing privilege for the ensuing year was sold at auction."

The town continued to sell the rights to take herring at its annual town meeting through 1906, but this changed the following year, delegating this task to the selectmen and thereby ending in Middleborough the uniquely New England practice of town meeting herring auctions.

Social Welfare

Recognized as a convenient and economical food source during the nineteenth century, local herring were adopted for the relief of the poor, providing a meal that was both inexpensive and nutritious. Herring allowed "the poorer classes at all times to have on hand an ample supply of healthy

Middleborough Poor Farm, circa 1905. For many years, residents of the previous home had their meals supplemented with Nemasket River herring. *Middleborough Public Library.*

food during the season, at a nominal cost" and may have enabled some not to seek the relief of the town, including "poor widows" who were furnished free herring as a means to ease their financial straits.

The practice of distributing herring to the community's poor and widowed was inaugurated in the eighteenth century and continued for several decades. In 1820, Middleborough listed some fifty-three residents as eligible for the benefit of free fish, including twenty-five widows. In that year, several thousand herring were distributed in amounts ranging from one hundred to four hundred at a cost of $46.25 to the town. In 1829, when the town formally established a municipal almshouse on Bridge Street, residents there began receiving free herring as a supplement to food grown on the farm.

Nineteenth-century Middleborough widows, in contrast, were provided with Nemasket herring at their own homes, although such largesse increasingly grew to be considered old-fashioned. Following the Middleborough town meeting vote of 1859 to provide two hundred free herring to those widows paying no tax, the *New Bedford Standard* remarked that the warrant article "read as quaintly as some of the proceedings in 'good old Colony times.' Middleboro' must be a good place for poor widows to move to—They all get two or three donation visits during the winter, a couple of hundred herrings free gratis and for nothing, and for aught we know a good pick for a second husband."

A surviving nineteenth-century anecdote relates to the widows' allotment. When one new selectman at the Middleborough annual town meeting had the temerity to ask just who exactly constituted a "poor widow," Sylvanus Hinckley quickly rose to his feet and preempted the moderator by chivalrously responding, "Mr. Moderator, any woman who has lost her husband is a poor widow, and there is your answer."

Most Middleborough widows did, in fact, avail themselves of this benefit. In the late 1860s at a time when the number of widows had increased as a result of the recent war, a quarter of each year's catch went to these women. So desirable was the widows' benefit regarded that some thought little of defrauding the town in order to take advantage of it. On record is at least one instance of an unnamed resident who "drew his widowed mother's allotment of herring [for] several years after her death before his trickery was discovered by the authorities."

Shortly after the Civil War, the practice of providing herring to widows was abandoned. Other communities followed suit. Both Rochester and Marion did away with the practice for Mattapoisett River herring some time before 1885, although the town of Mattapoisett continued to oblige its widows. Eventually, even that town eliminated the right altogether when the privilege there began being auctioned to private individuals.

Besides the poor and widows, herring were also used to relieve the hardship of others on rare occasions. During the Civil War, wives of soldiers on active duty were provided with two hundred herring. Somewhat disturbingly for the superstitious, this was the same entitlement that widows received.

Protecting the Victorian Fishery

Because the Nemasket fishery remained a valuable economic asset and played an important role in municipal tax relief and social welfare programs, it continued to require protection throughout the nineteenth century. It was to the fish wardens that this task fell.

Fish wardens were paid for their services, receiving $1.50 for every twenty-four hours of service in 1801. Apparently, some question arose as to whether wardens were truthfully reporting their time, for in 1803, it was voted that they attest under oath concerning the number of hours spent monitoring the local fishery. At that time, their compensation was reduced to $1.00 for

REPORT OF FISH WARDENS.

HERRING ACCOUNT.

Received from sale of herrings	$275 00	
Received from towns on Taunton River for inspection at East Taunton,	30 00	
		$305 00

Orders drawn :		
Town of Lakeville for 1895 and 1896,	$49 17	
H. L. Leonard	10 00	
Allen B. Thomas	130 00	
Katharine A. Sproat . . .	10 00	
A. T. Savery	5 00	
E. T. Lincoln	5 00	
C. W. Kingman	7 75	
E. F. Witham	8 00	
		224 92

To divide with Lakeville		$80 08

ALBERT T. SAVERY,
EDWIN F. WITHAM,
CHARLES W. KINGMAN,
Fish Wardens of the Town of Middleborough, Mass.

First annual report of Middleborough's fish wardens, 1897. *Author's collection.*

each twenty-four hours, and this amount did not increase substantially over the course of the century.

Much of the work conducted by these wardens involved the apprehension of poachers, a lingering problem. Poaching remained the most serious problem for nineteenth-century wardens, as it had the potential to greatly compromise a run. As early as 1823, poaching was cited as the reason for the failure of the Mattapoisett River fishery to reach its full financial potential. For those individuals (including Elisha Tinkham and Ebenezer Ellis in 1784, James Sprout and Ebenezer Finney in 1795 and Ebenezer Pratt in 1797) caught taking small numbers of fish from the Nemasket River, fines of a pound were levied. Transgressors on a grander scale—who took alewives through the use of seines—faced much stiffer fines. On October 30, 1778, the town received eighteen pounds and seven shillings from a Taunton resident named Williams for unauthorized seining in the river there, while an even larger sum of ninety pounds was received through the town's agent, Silas Wood, on May 19, 1781, the aggregate of a number of fines levied "on account [of] seining alewives contrary to law in Taunton Great River." More than a century later, when widespread reports of poaching at Muttock were made in 1902, they were treated seriously. Later, signs were erected for which the town paid the Clark & Cole Box Company three dollars in 1911, doubtless outlining the restrictions at each site and warning off potential violators.

In addition to deterring poaching, Middleborough's fish wardens were equally focused on protecting the fish from harassment. An 1863 Massachusetts law provided punishment for those convicted of molesting the fish by "beating upon the ground" or "doing any act, whatsoever, whereby the fish shall be disturbed, hindered, driven, or delayed on their passage up said river, from the first of March to the tenth of June." Fines were set between five and twenty dollars and imprisonment at sixty days. The town had no compunction about prosecuting violators of these fish laws, paying attorney John C. Sullivan of Middleborough fifteen dollars in 1879 to do so.

The role of the fish warden was often difficult and frustrating, thanks to violators who sought to circumvent the fishery regulations and annoy the wardens. "The officers were subjected to great indignities, such as the ingenious and reckless company of enterprising youth could suggest." However, the powers of the wardens were extensive and included the legal right to retain property used in herring poaching. By 1897, the Middleborough fish wardens' role had grown to such an extent that a level of public accountability was expected, and that year marked the first in which the three elected fish wardens prepared an annual report of their activities.

COOPERATION AND CONFLICT BEYOND NEMASKET

As demonstrated by the oversight provided by Middleborough's fish wardens, the town was keen to protect its fishery from any local threats. As the nineteenth century wore on, potential harm to the Nemasket fishery was seen as coming evermore from outside the community, and the town became increasingly involved in working with other towns to protect its local fishery.

The first local cooperative measure occurred in 1853, when the town of Lakeville was set off from Middleborough, a portion of the corporate boundary between the two being established as the middle stream of the Nemasket River. At that time, it was agreed that the new town would retain 15 percent of the Nemasket herring fishery. Such an arrangement was not unique. In 1739, when the town of Wareham was set off from Rochester, "a weir was reserved for Rochester on the Weweantit River," and when Marion and Mattapoisett similarly left, they were each permitted an inspector for the Mattaposiett River. Hanson residents likewise retained rights in their former fishery when they separated from Pembroke in early 1820.

Following the 1853 division of the Nemasket fishery, Middleborough initially conducted the business, including harvesting the fish, without assistance from Lakeville. Eventually, however, the town appears to have tired of this arrangement, as indicated by the somewhat exasperated tone of the town meeting of March 1856 that "voted to invite Lakeville to take their share of herrings at the fishing places and dispose of them themselves." Nonetheless, the two communities enjoyed an amicable relationship regarding their common herring fishery in contrast to communities elsewhere, where both privileges and occasional animosity were shared. During the spring of 1906, Marion's Inspector of Herring Andrew N. Gifford made news when he chastized a number of Mattapoisett boys for fishing at the Sippican River weir. Gifford's counterpart in Mattapoisett, Ezra N. Brigham, took umbrage and indicated that Gifford would "run the risk of receiving a coat of tar and feathers or a ducking in the river" were he ever again to interfere with the Mattapoisett privilege.

Despite the concession to Lakeville at the time of that town's incorporation, Middleborough historically had been reluctant to relinquish control over its fishery and was particularly vigilant regarding developments downstream on the Taunton River for fear they might compromise the Nemasket fishery. In March 1765, when the Massachusetts legislature passed a law limiting the taking of herring, Middleborough interests were successful in framing the legislation so that seining on the Taunton River could commence only "after it shall be known that alewives have been taken at Middleborough, in the spring of the year, annually."

Taunton felt itself disadvantaged by the law and in 1773 petitioned for permission to operate additional seines. Its petition reflects an underlying resentment toward Middleborough, which was becoming well known for its herring:

> [T]*he Alewives passing by said Mill River proceed up to Middleboro & other Towns where they are taken by Scoop Netts with great Ease, and at Middleboro in great plenty, so that for many years past the chief of the Alewives that have passed up Taunton Great River which have been taken have been taken at Middleboro.* [The inhabitants of Taunton are] *entitled by nature to at least as great a proportion of Alewives as the Inhabitants of Middleboro & other Towns and on some accounts more, they think it hard to have the Alewives pass their Doors thro' the heart of their Town without such a regulation of the taking of them that they may get their proportion of those proper to be taken, & to be obligated to undergo the expence & trouble of going to Middleboro to buy the very Fish that went by their Doors.*

For its part, Middleborough remained sceptical of Taunton's willingness to comply with laws regulating the herring fishery and so began sending representatives to monitor the run along the Taunton River at this time. In the early 1780s, Joseph Lovell of Middleborough was paid thirty pounds "for his taking care of the fish down Taunton River," and Joseph Tupper received sixty pounds from the town for "inspection of the river in Bristol County."

Increasingly concerned about potential overfishing, Middleborough began to oppose more actively efforts by downriver communities to secure or extend fishing rights along the Taunton. Between 1802 and 1804, Middleborough challenged the efforts of neighboring Raynham to acquire a fishing place on the Taunton River, fearing the impact it might have on the number of local herring. In 1816, the town secured further restrictions on the Taunton River fishery when the legislature made it illegal to catch or destroy herring within the limits of Middleborough or in the Taunton River in Middleborough, Bridgewater or Raynham, except at certain specified locales, and authorized the town to appoint herring agents for the enforcement of this act. Nine years later, in 1825, the town sought repeal of the previous year's law permitting Bridgewater to sweep the Taunton River with a seine nearly 250 feet wide four days a week. In 1827, Middleborough petitioned for a reduction in the number of fishing places along the Taunton River and for further restriction on the times for taking fish. The town supported the 1843 passage of *An Act to Regulate the Fishery in Taunton Great River* that strictly regulated the fisheries of Somerset, Freetown, Fall River, Berkley, Dighton and Taunton, limiting the times, locations and means in which fishing could be conducted, and in early 1844, it petitioned that the act be continued.

REGULATING THE FISHERIES: THE 1855 AND 1863 LAWS

Despite these laws, increasing regulation of the Taunton and Nemasket River fisheries was not always welcome in Middleborough, particularly following 1855, when the state legislature authorized a number of seining privileges on the middle and lower Taunton River. Middleborough residents were outraged by this concession and sought to have the law overturned. Stillman Pratt, editor of the *Namasket Gazette*, wrote a scathing and potentially libelous commentary attacking the downriver interests he saw responsible for the act's passage:

> *We understand that our friends below were exceedingly kind to the Legislative Committee, furnishing them with fat shad in abundance....*
> *We do not affirm that the Legislative Committee on fisheries were bribed to curtail our privileges and increase those below us, but we do know that presents have a most emollient tendency.*

Like Middleborough residents, inhabitants of Taunton, Raynham and Berkley were displeased with the 1855 law regulating the Taunton River fishery. Feeling shortchanged by the law's provisions, they petitioned in October 1859 to have a portion of the act repealed so that no distinction would be made between the middle portion of the river and the Nemasket River regarding on what days herring could legally be taken.

Middleborough's dissatisfaction with the 1855 law merely strengthened the town's resolve to protect the Nemasket fishery, a mood reflected in an 1857 town meeting decision to "deny the right of any other corporation in the management of the herring interest, or its proceeds, within the town." At the time, it was further voted to instruct Middleborough's legislative representative at Boston "to

Taunton River fishermen, circa 1900. Unlike their Middleborough counterparts, Taunton fishermen felt disadvantaged by restrictions on the local herring fishery and sought their repeal. *Author's collection.*

procure a law leaving the management of the herring interest within the limits of the town with the inhabitants of the same." Although the vote had no practical impact, it served as an unequivocal indication of the community's resolve not only to exclude outsiders from the Nemasket fishery but also to protect the fishery through extending its control over a portion of the Taunton River.

Perhaps no clearer demonstration of the lengths to which Middleborough was willing to go to protect its fishery was an incident from the spring of 1859, when Middleborough fish wardens took decisive action against one downriver community:

> *The dangers of war are not confined to Europe. The New Bedford (Mass.) Standard informs us that the annual herring war is raging vigorously between Taunton and Berkley. The people of the latter town having a particularly advantageous situation for their weirs, caught all the herring that came along, leaving hardly a struggle for the good people of Taunton and the widows of Middleboro. There was like to have been a genuine herring famine in these unfortunate towns. And so falling back upon their*

reserved rights as herring eaters, and ignorant, we presume, of the recent decision of the Supreme Court in regard to the abatement of nuisances, the fish wardens of Middleboro' and Raynham made a combined attack upon the Berkley waters, and captured and carried off the seines. The crop of lawsuits will be larger than the crop of herrings.

More constructively, during the late autumn of 1862, Middleborough and Lakeville successfully appealed to the state legislature for a revision of the 1855 law, and in 1863, Middleborough's authority to oversee the Nemasket River fishery in its entirety was confirmed and the town's oversight extended to include that portion of the Taunton River running between the mouth of the Nemasket River and a mile and a half downstream from the East Taunton dam. The expense of monitoring was to be borne by the several communities located downstream on the Taunton River, which were required to reimburse Middleborough for patrolling the river.

Not surprisingly, the Taunton River communities opposed this new law. In early 1868, the City of Fall River looked to opt out of the legislation by forfeiting its rights in the fishery, and in 1872, when one Berkley commentator indicated that "legislation to regulate these fisheries has not been satisfactory to all," he revealed the lingering discontent that was rife in that town.

For its part, Middleborough saw its efforts to protect the Nemasket fishery finally codified into law, and the community remained ever vigilant to prevent any alteration of the 1863 act. In 1911, when Senator Charles H. Chase of Bristol County introduced a bill that would have stripped Middleborough of its patrol and supervisory functions and transferred them to the commonwealth on the grounds that the state could provide better supervision of the Taunton River, the bill was vehemently opposed in Middleborough. Judge Nathan Washburn bluntly dismissed the bill as "just so much more deadwood and useless legislation" and, along with Selectman Haskins of Middleborough, detailed the responsible oversight the town had provided, as well as cited its efforts at East Taunton. Nothing came of the bill.

MONITORING EAST TAUNTON

With the extension of Middleborough's authority over a portion of the Taunton River came a formalization of its role at East Taunton, where the

East Taunton dam and gatehouse, early 1900s. The fishladder described as "an inclined plane" is just out of view to the left. *Author's collection.*

sole dam on the Taunton River below the Nemasket was located. Because this dam posed a potential threat to the passage of the herring upstream to the Nemasket River, its maintenance—particularly that of its fishway—became a focus for Middleborough's fish wardens, who remained a presence at East Taunton for decades.

The 1813 construction of the original stone dam at Squawbetty (now East Taunton) was authorized with the proviso that "a convenient way… be constructed and kept open according to law, for the passage of such fish as usually pass up the same river in their proper season." Ten years later, in 1823, when the Squawbetty dam was relocated downstream to present-day Old Colony Avenue, where a fall of nine feet of water was developed sufficient to construct an ironworks, the owners were required to maintain this passage.

The Squawbetty fishway functioned effectively for half a century, but in 1861, the small numbers of fish in the Nemasket River raised concerns about its efficacy. Limited numbers of herring above the dam at East Taunton were particularly alarming, as that community had long been known for great numbers of the fish. "There is a place at East Taunton where the river at times becomes so crowded with herring as almost to suggest the possibility of walking across it on the silvery-gray backs of its animated pavement." Upon invesitigation, it was found that "not more than half the fish that reached [the East Taunton] dam were able to ascend any further." Middleborough voted to notify the owners of the dam that the town "shall depend upon them to get the fish by the dam." Part of the problem may have been with the fishladder itself, described later in 1870 as "an inclined plane, down which water rushes with great force."

Two years later, when its authority over this portion of the Taunton River was established by law, Middleborough adopted a more proactive role at East Taunton through the agency of its herring inspectors. These men, appointed by the Middleborough selectmen, made frequent visits and were generally present for a number of days during the annual spring run, keeping a watchful eye on the dam to see that nothing hindered the upstream progress of the fish.

Inspectors regularly patrolled the river at East Taunton looking for malefactors, an endeavor in which they met with frequent success. As at Middleborough, the work of the inspectors at East Taunton involved the apprehension of both poachers and individuals who harassed the fish, actions illegal under terms of the law. To aid in this task, East Taunton inspectors typically carried the title of fish warden, and this empowered them to seize the property of transgressors.

Harrassment of the fish was frequent. Boys were noted for tormenting the herring on their upstream migration by throwing rocks at the fish, which proved easy targets as they made their way through the ladder. In May 1859, upon a complaint filed by the Middleborough fish wardens, three individuals were convicted in Taunton for throwing rocks into the river "thereby disturbing the fish known as shad and alewives." Each was fined ten dollars and sentenced to one day in jail. The 1863 law strengthened the ability of Middleborough's East Taunton inspector to protect the fish by providing authority to prosecute "[a]ny person who shall beat upon the ground, or do any act whatsoever whereby said fish in said rivers shall be disturbed, driven, destroyed or delayed, from the first day of March to the tenth day of June in each year."

Of all activities at East Taunton, however, it was poaching that was most common and most notorious, as noted by Thomas Weston: "After sunset, men, as well as boys, had their secret hiding-places to catch the fish secretly, and boasted of their thefts afterwards and of their escape from the fish wardens.... It was thought the best of sport, and the convictions were so infrequent that these escapades were regarded with special zest by a large

James A. Burgess. The *Middleboro Gazette* called Burgess "that staunch champion of the rights of alewives and oracle upon the subject of local fisheries." *Author's collection.*

number of people who would probably not care to have their names known."

In May 1864, upon a complaint filed by Middleborough fish warden Reuben K. Simmons, two Squawbetty residents—John McGrath and Timothy Shea—were convicted in Taunton Police Court of illegally capturing herring and fined ten dollars. Nonetheless, poaching remained an openly acknowledged fact. In April 1875, the *Taunton Daily Gazette* reported what Middleborough fish wardens already knew: "In spite of precautions it is said a large number of people at East Taunton and Raynham have been feeding on herring surreptitiously taken from the river."

According to James Burgess, considerable poaching occurred in the vicinity of the Squawbetty dam, and he once said of the perpetrators:

> *They seemed to think that the right that was given by the whites to the old squaw, Betty, from whom the land that Taunton now occupies was purchased, still belonged to them, as though they were their descendants. And today any person born in Squawbetty thinks that if he can get the early herring in the same old way they are more palatable to the taste.*

Naturally, free herring were more palatable than those that had to be purchased, and poaching continued unabated. So profitable was the activity that rival gangs sprouted up at East Taunton to take advantage of the situation. Burgess was informed in the mid-1870s by the superintendent of the Old Colony Works, which stood alongside the Taunton River at Squawbetty, "that there were apparently two regularly organized gangs of herring thieves." Consequently, "the fish wardens are keeping a sharp eye on the herring and scaly depredators find but poor chances to abstract any of them." During one episode, Burgess, accompanied by Middleborough and Raynham fish wardens, ambushed herring thieves who had come up the Taunton in the early morning with the intention of dragging the river for the fish. "They fought the case clear to the Superior court. I won my case, and that closed the organized stealing there," reported Burgess.

Despite the law, unorganized poaching continued at East Taunton, driven by the fact that it remained a lucrative undertaking—so much so that even at least one local official fell afoul of the law. In the 1870s, Burgess unmasked collusion between an appointed herring agent and a gang of East Taunton poachers that emptied the fish boxes at Squawbetty early in the season. The man was summarily fired and the poaching operation effectively shut down. In 1879, the fish wardens caught a number of violators, who were brought to court, where they were prosecuted by attorney John C. Sullivan of Middleborough on behalf of the town. Sullivan was successful and in one engagement during May won a sum of seventy-six dollars for the town.

Unlike Burgess, not all of Middleborough's agents at East Taunton were compensated, and some had to assume responsibility for financial losses sustained in the conduct of their duties. While on duty in late April 1861 as Middleborough's herring agent, George W. Wood had "his horse…thrust into the [Taunton] river and drowned." Although Wood was engaged in official work for the town, the Middleborough annual town meeting of 1862 rather meanly denied his request of $152 compensation for his horse.

Perhaps the reason for the town's stinginess was the already considerable expense of monitoring the East Taunton dam, a burden that until 1863 fell exclusively on Middleborough, which spent more than $200.00 each year on the task. So hefty was the expense that in 1855, it was recommended that the cost of monitoring alewives at Squawbetty hitherto financed by Middleborough "be in future averaged on all the towns on the banks of these rivers as a matter of simple equity." In 1863, it was made law that Middleborough would receive $10.00 for each of the thirteen seining privileges on the Taunton River below Nemasket to help defray the expense of monitoring the fishery and the East Taunton dam, income that was sorely needed with the formalization of Middleborough's role at East Taunton. In addition to the wages of inspectors, herring agents and night watchmen, Middleborough frequently had to pay to board its inspectors at East Taunton. In 1867, widow Laura W. King, who resided in East Taunton near the Squawbetty bridge, was paid $104.00 and again in 1871 was paid $60 "for boarding watchers." J.S. King was paid $40.00 the following year, and in 1879, $36.28 was paid to F.A. Bosworth for "board at East Taunton." Not inconsiderable, either, was the cost of transporting the

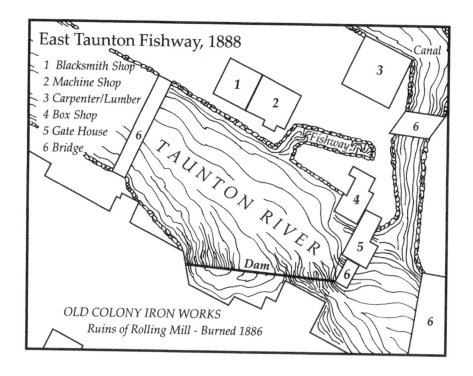

East Taunton Fishway, 1888

1 Blacksmith Shop
2 Machine Shop
3 Carpenter/Lumber
4 Box Shop
5 Gate House
6 Bridge

Canal

TAUNTON RIVER

Fishway

Dam

OLD COLONY IRON WORKS
Ruins of Rolling Mill - Burned 1886

Taunton River seining privilege, circa 1910. Such privileges were responsible, in part, for financing Middleborough's oversight of the river. *Author's collection.*

inspectors to East Taunton, for which purpose I.G. Grew was paid $10.00 for "car fare" in 1872.

Revenues generated from each of the thirteen Taunton River seining privileges fluctuated greatly, and the downriver communities were sometimes negligent in their payments to Middleborough; this only exacerbated the Nemasket fishery's poor financial situation. It appears that no money was forthcoming in 1900 from downriver towns, and in subsequent years, receipts fluctuated sharply. Meanwhile, Middleborough's wages paid to East Taunton inspectors remained unchanged at $130.

Middleborough retained its inspectorate at East Taunton through the first quarter of the twentieth century. The position (which in 1915 carried the title of fish warden and inspector of alewives at East Taunton) was no longer filled following 1925. With the collapse of the East Taunton dam in 1930, it was no longer considered critical for Middleborough to monitor the site.

ACUSHNET RIVER FISHERY

Westward toward the Taunton River was not the only direction in which Middleborough cast a cautious eye in protecting its herring during the nineteenth century. Proposals to reopen a fishery in the Acushnet River by tapping into the southern end of the Middleborough ponds greatly unnerved Middleborough residents. Particularly upsetting was the fact that the Acushnet River fishery had been destroyed largely through mismanagement and neglect.

The engineering of natural watercourses to create or expand herring runs was not unknown in southeastern Massachusetts. Nantucket's Madaket Ditch, constructed in 1665, was probably the earliest regional instance of a man-made channel dug to promote a herring fishery. Closer to Middleborough was the example of Snipatuit Pond in Rochester, where in the late 1700s the Town of Rochester opened an artificial channel "at great labour and expense" between Snipatuit and the Mattapoisett River to encourage "the yearly passage of vast shoals of herrings into Snipatuit." In 1855, Thoreau, who not surprisingly was familiar with this alteration, urged his friend Daniel Ricketson of New Bedford to visit the pond.

Encouraged by the success of the Snipatuit Canal, work commenced in the 1830s on a similar channel linking the Acushnet River with Little

Great Quitticus Pond, circa 1910. In 1907, it was remarked that "formerly large quantities of alewives went into it...but very few pass now." *Author's collection.*

Quitticus Pond as a means to revive that river's herring fishery. In 1823, Reverend Dr. Abraham Holmes, writing at Rochester of Great and Little Quitticus Ponds, noted that "formerly large quantities of alewives went into [them] through a small brook from Assawompsett Pond, but very few pass now." The construction of the canal was either never completed or was not successful, for no further mention is made of it. Similar failure greeted an effort by the town of Rochester to open additional spawning grounds by cutting a canal between the Sippican River and Mary's Pond at a cost of $100 in order to induce herring upstream. The result was curtly noted: "No success attended the attempt."

These various projects had been virtually forgotten when the Little Quitticus proposal was resurrected in the 1850s as a means to revitalize the Acushnet River fishery, which had once operated at New Bedford and Fairhaven but had declined through gross mismanagement and neglect, as well as the lack of suitable spawning grounds for the herring. In early 1857, New Bedford, Fairhaven, Freetown, Lakeville and Rochester jointly proposed the construction of a canal linking the east branch of the Acushnet River with Little Quitticus Pond. The towns requested authorization to "stop, close and fill up all the other channels and outlets" from Little Quitticus, including that into Great Quitticus, and to erect a dam at Stone Brook to prevent the waters of Long Pond from backflowing the canal, with the intention that the pond's waters would "forever hereafter" be discharged into the Acushnet River. On April 11, 1857, Thoreau, perhaps apprised of the proposal by Ricketson, made a notation in his journal that "[m]any inhabitants of the neighborhood of the ponds in Lakeville, Freetown, Fairhaven, etc., have petitioned the legislature for permission to connect Little Quitticus Pond with the Acushnet River by digging, so that the herring can come up into it."

In a *New Bedford Standard* editorial directed at a Taunton audience (but which applied equally well to readers in Middleborough), Tauntonians were advised to take heed of the proposal:

> *Our Taunton friends will take warning, and see that their herring reputation be not lost, for if the outlet of Middleborough Ponds be turned this way, the water never can again be induced to take the old circuitous route by Taunton Green to the sea.*

Middleborough opposed the plan, arguing that restriction of the natural outflow from Little Quitticus would lower the level of the Nemasket River, thereby impeding industrial development in Middleborough and compromising the Nemasket fishery. Others warned of possible natural disasters that would ensue as a result of altering the flow of the pond, including *Namasket Gazette* editor Stillman Pratt, who cited the 1810 flood in Vermont's Barton River Valley that resulted when residents attempted to redirect the flow of Runway Pond from the Lamoille River into the Barton River. Within fifteen minutes of the diversion of its flow, Runway Pond had been drained, and a wall of water some sixty to seventy feet high and three hundred feet wide had crashed down the valley, felling trees, moving boulders and destroying homes before reaching Lake Memphremagog twenty-seven miles away.

The *New Bedford Mercury* accused Middleborough of being "small" and petty to "contend against the removal of one twentieth of the waters of the Nemasket River," and in response, Pratt clarified the community's position: "The principle and the extent to which it may be carried, is the thing complained of." Opponents of the plan, like "S.H.," who wrote to the *Namasket Gazette* in May 1857, suggested that interest in creating a remunerative fishery for the south coast towns could better be satisfied by stocking Quitticus with "foreign fish much more valuable than herrings." S.H. also suggested that the city's desire for access to the pond had little to do with herring and more to do with gaining access to the pond as a municipal water supply: "Herrings are the specious plea for securing a mighty advantage to New Bedford." This contention was later corroborated to an extent when the city constructed a 411-acre municipal water supply reservoir at the headwaters of the Acushnet River at Long Plain in 1869.

Although the Acushnet River herring fishery proposal was eventually defeated in the legislature, the lessons learned from it, the strategies suggested and the resentment engendered would resurface at the close of the century when the cities of Bristol County sought to tap the Middleborough ponds as a municipal water supply. The result was long years of litigation as affected towns and industries sought damages for the loss of water, in the process fostering a lingering animosity between water-rich and water-needy communities in southeastern Massachusetts.

THE CULTURE OF THE
NEMASKET FISHERY, 1801–1900

OLD AND NEW USES FOR HERRING

Although commercial substitutes had long since replaced herring as a fertilizing agent by the mid-nineteenth century, the abundance of the fish during the spring runs was at times put to agricultural use, excess herring being strewn about tilled fields and plowed into the soil. James A. Burgess recalled this practice from his youth at Muttock, where at times "there were so many fish that they could not be disposed of and they had to be thrown upon the ground and ploughed in for dressing." In 1862, herring were particularly abundant in the Taunton and Acushnet Rivers, and estimates gauged the run as the most populated in fifteen years. So heavy was the run that the owner of the East Taunton privilege "was compelled to load the herrings into his carts and use them for the purpose of enriching the land. Several thousand were thus disposed of." Middleborough farmers in May 1877 similarly were said to be "trying the quality of herring as a fertilizer for land," buying the fish at the low price of 10 cents per 100.

Local farmers were encouraged to employ herring as fertilizer by agricultural and other journals, which touted the efficacy of fish in enriching the land. In 1801, the *Columbian Courier* of New Bedford, which reached households in Middleborough, cited the value of the practice

Fishing pool below the East Taunton dam, circa 1930. For generations, the site was a popular location for taking herring. *Author's collection.*

where in one case in England thirty-two thousand fish were plowed into an acre of land, with startling results. Likewise, the prestigious *DeBow's Review* in 1866 said of the menhaden (which is similar to the herring) that "a single fish of medium size has been considered equal, as fertilizer, to a shovel-full of barn-yard manure."

Herring were also increasingly recognized for their value as bait. Reverend Holmes in Rochester recommended the use of herring fry taken in the fall for catching white perch, stating that it was "the only bait which can be used with success."

Despite these uses, the role of herring as a food source gradually overtook its agricultural importance. Herring were eaten and were a decidedly acquired taste according to those who ate them. Although the Plymouth *Old Colony Memorial* termed the fish a "bony and toothsome fish" and in 1877 the *Middleboro Gazette* advised "get your boning mills ready" preparatory to the run of that year, herring would long afterward be considered "tasty local favorites." The editor of the *Gazette* in 1873, upon viewing the thousands of fish moving upstream, saw only the "promise of many a rasher for an evening meal," and his successor nearly a century later would share the same fondness for the fish, impatiently awaiting their springtime arrival.

Catching Herring

One of the iconic images of nineteenth-century Middleborough is the local herring catcher, scoop net in hand, surrounded by a multitude of barrels and crates. Worth quoting at some length is the following description, carried in 1897 in the *New York Times*, detailing the manner in which herring were harvested at that time at Muttock:

> *At the lower junction of the canal* [used as a fishway] *and river a net is placed across the latter, thus forcing the fish to continue their course up the canal. On the off days, when all dams and obstructions are removed and the net is taken away, the fish, perhaps through fright or attracted by the quiet waters, swim into the space between the junction and the falls and accumulate into the thousands. There are sometimes 30,000 or 40,000 fish preserved here, and when the day for catching them again comes around, the net is pulled down, and the fish remain there—waiting for a rise in the market.*
>
> *About 150 feet above the lower junction a temporary dam is thrown across the canal, the lower part of which is made of pine planks, surmounted*

Muttock herring harvest, late nineteenth century. The photograph demonstrates how fish were captured in enormous scoop nets and deposited into waiting barrels. *Author's collection.*

by a moveable screen of coarse wire bars, through which the water passes and falls about three feet. Further down the canal a huge coverless wooden box is sunk, extending across the waterway. It is plainly visible, as the water is shallow and clear. The fish coming up the stream swim over the box and on until they reach the dam; there they are thrown back by the force of the water, and many of them jump out upon land in their endeavor to scale the fall. Others resort to deep pools sheltered by rocks. Here they hide and are sometimes successful in making their escape by remaining during the fishing days.

When a sufficient number of fish have gathered between the box and the dam, a net is stretched across, below the box, and the fishermen don high rubber boots, and, starting at the dam, walk downstream, driving the fish before them into the sunken box. Here other men are stationed with huge landing nets; they scoop up the fish, throw them upon the platform, from which they are taken and packed into barrels partly filled with ice, and sent away to market. Nine catches are made daily, the average catch being eighty barrels.

The present manner of taking the fish has been in vogue almost from the settlement of the town.

Retailing Herring

Middleborough's nineteenth-century dining tables were supplied with herring by the several groceries and fish markets that sprung into operation during the century. In February 1855, Southworth Loring opened a fish market in the basement of the American Building on South Main Street, and among the products that he advertised was herring. Stillman Pratt of the *Namasket Gazette* sampled some of Loring's fish and told his readers about them in the February 9 edition. "Having tried some of his fish we are prepared to recommend them to our neighbors, and we welcome this new appendage to the convenience of our village." The next year, following the annual run, Loring & Cushing was advertising smoked and unsmoked herring for 60 cents per 100. J. Herbert Cushing would long be noted as a fish dealer in Middleborough, and among his stock were always herring, fresh not only from the Nemasket but from elsewhere as well. In March 1895, he was noted as "distributing some fine" Martha's Vineyard herring to his customers. Grocer E. Tinkham also sold the fish, advertising in the pages of the *Namasket Gazette* in June 1853 that "[t]he Herrings have sent a full delegation into our Namasket river this year and a lot of the best of them are for sale at Mr. E. Tinkham's Grocery and Grain Store."

Foremost among these men, however, was Cushing's brother-in-law, David A. Tucker, who operated a fish business "on an extensive scale" until about 1885, when he retired. Locally, Tucker was noted as the "Herring King": "He formerly bought practically all the herring which swam into the Taunton and Nemasket rivers, as well as buying up the fishing rights on several other streams about here." Tucker's herring sales allowed him to expand his financial interests into local real estate, and he was the owner of the American Building on South Main Street, where his market was located. Tucker was eventually able to become "one of the most prosperous men of the community by virtue of the alewives."

Tucker became so associated with retailing herring that the *Middleboro Gazette* could refer to him solely by his Christian name: "David has fresh herrings, good; now we can live again." In 1872, Tucker was the first to offer herring that season, "and they went quick at about four cents a piece." In years such as 1875, when he was unable to sate Middleborough's demand for early herring with local fish, Tucker procured herring from Martha's Vineyard to supply his customers.

For a quarter, individuals in about 1900 could purchase a stick of a dozen smoked herring from where they hung in a frame in the Cobb, Bates & Yerxa store at the Four Corners. Grace Clark, whose grandfather managed the store, recalled, "Grandpa used to tell the boys, Eleazer, Bay, Horace, Johnnie and 'Himself' whenever they sold a stick of herring to always take the top rows on the frame that held, as I remember, ten or twelve sticks. He said the older men knew without telling them." As late as 1916, grocers Lucas and Thomas, located in the Thatcher Block on Center Street at Thatcher's Row, were advertising "Cape Cod Smoked Herrings" at 30 cents per dozen.

The first herring of the season normally fetched the highest prices, so eager were buyers for fresh fish. The first one captured in 1856 at Dighton sold for $1.50. Typically, however, prices declined quickly over the course of the season as the fish became more readily available.

The sale of herring, like its processing and preparation, was accompanied by less than pleasant odors. Edith Austin Holton wrote that one "could always tell where a herring seller lived by the smell of his back yard. There has never been an effluvium like it since and it is just as well if there never is again." Novelist Joseph C. Lincoln also noted the prevalence of the herring

Cobb, Bates & Yerxa, Middleborough Four Corners, circa 1900. Smoked herring were sold from racks like that seen to the left. *Author's collection.*

Herring house from *Cape Cod Folks* (1904 edition). The powerful smell emanating from such buildings was characteristic of fish processing sites like Muttock. *Author's collection.*

smell where sticks of the fish were hung to dry: "That is one characteristic of a herring which salting or drying does not remove, but rather accentuates—the smell…. Call him an alewife if you will, but, like the rose by another name, it does not change his aroma." The smell was all-pervasive:

> *Against the rafters of practically every barn, and in many sheds and outbuildings, those sticks of herring used to hang. You could smell them before you opened the door. During the summer they were often hung out of doors in the sunshine, festooning the eaves of barns and sheds. You could smell them there, too.*

Locally, this seems to have been less of a problem than it was at Taunton, where herring peddlers set up shop at the southeast corner of the Green, vexing many nearby shopkeepers and passersby with their pungent-smelling goods.

Preparing Herring

While it was known for herring to be eaten fresh (novelist Joseph C. Lincoln recalled his grandmother's declaration, "I do relish a nice fresh herring with my breakfast"), the majority were dressed or preserved through salting and smoking or, less commonly, pickling in brine. Middleborough's Clint Clark reported having once met a gentleman whose mother pickled the fish in jars. One of the advantages of the herring as a food was the fact that in the days prior to refrigeration, it was easily preserved.

Fresh herring were seen at the start of the season and lasted only for a short time, whereafter only corned (salted) herring and smoked herring were available. "By June if you wanted herring you had to take them smoked or go without." Fresh and smoked herring were the varieties preferred at Middleborough, as noted by a resident who remarked that in the early 1900s there were no corned herring in town, as "everyone baked them fresh." In nearby Pembroke, residents preferred the opposite. "Certainly at most Pembroke tables a fresh herring two days corned, and served piping hot with salt, pepper, vinegar, and leisure to eat it, is esteemed the only right supper for a damp, chilly spring evening."

In addition to the fish themselves, herring roe was considered a delicacy favored by many. "Have You Tried It?" a 1909 advertisement for the product

Packing herring roe, 1940s. Canned river herring roe (often fried into cakes) was widely marketed until the second half of the twentieth century. *Author's collection.*

To broil Herrings. **Scale and gut them, cut off their heads, wash them clean, dry them in a cloth, flour and broil them; take the heads and mash them, boil them in small-beer or ale, with a little whole pepper and an onion. Let it boil a quarter of an hour, strain it; thicken it with butter and flour, and a good deal of mustard. Lay the fish in a dish, and pour the sauce into a bason; or plain melted butter and mustard.**

Recipe for boiled herring from Susanna Carter's *The Frugal Housewife* (1803). It incorporated all parts of the fish, including the head. *Author's collection.*

from the Middleboro Fish Market asked. "Delights the palate of the epicure. Nothing more delicious can grace the table." Although Grace Clark disliked the fish because of its numerous bones, she "did like the brown streak of fat on the inside of the skin, and the roe." To determine whether fish contained roe, *Middleboro Gazette* editor Lorenzo Wood suggested gently squeezing freshly caught fish on the underside. Those that excreted a reddish material contained roe; those that excreted white did not.

Cookbooks throughout the era featured "receipts," or recipes, for the preparation of river herring. Susanna Carter's 1803 *The Frugal Housewife* taught readers how to fry and broil herrings. Mrs. N.K.M. Lee, in her *The Cook's Own Cook Book*, published in Boston in 1832, listed recipes for boiled, broiled, baked and fried herring and included tips for identifying fresh herring: "Of the fresh herring the scales are bright if good, the eyes are full, and the gills red, the fish should also be stiff." Eventually, however, the increasing lack of such recipes in cookbooks published as the nineteenth century wore on demonstrated herring's dwindling popularity at the expense of other freshwater fish, including shad, for which recipes were generally included. For instance, while Fannie Farmer's influential *The Boston Cooking-School Cookbook* (1896) included recipes for planked shad and fried, broiled and baked shad roe (the latter with tomato sauce), no recipes were included for river herring.

In preparing the fish for cooking, salted herring needed to be soaked beforehand to remove as much brine as possible, and these were generally broiled, although on occasion they were cut into pieces and eaten raw. Many cooks at the time had strongly held views regarding the dressing of salt herring, a hotly debated topic. J.M. Sanderson of the Franklin House in Philadelphia argued that "the New England mode of dressing salt fish is an excellent one, and ought to be generally adopted. Keep the fish many hours (at least seven or eight) in scalding hot water, which must never be suffered to boil." New Englanders, however, disagreed. Maria Eliza Rundell recommended pouring "some boiling small beer over them to soak half an hour," while Mrs. N.K.M. Lee favored either simply soaking the fish in cold water or soaking in cold water "a sufficient time" followed by two hours in milk.

The first step in preparing smoked herring—sometimes called red herring, as the smoking process turned them a golden color—was to remove them from the stick, a task impossible to accomplish "without getting the flavor transferred to your hands." The fish were then washed but were never soaked, which would have diminished the flavor.

Once prepared, the corned or smoked herring were then laid in a roasting pan to cook, frequently on brown paper. The pan in which the fish were cooked eventually became devoted to the purpose, for obvious reasons:

> *Any pan that started life as a herring roaster, you must realize, was due to stay that way through its entire career. It took no caste system to enforce this. No matter how vigorously a pan was scoured with wood ashes, soap and scalding water, it could be recognized ever after the event as a herring pan. Herring pans, therefore, always retired humbly to the wood shed between the acts.*

Grace Clark concurred, recalling that her mother had a pan devoted to the fish. "Never, no never, would any food be cooked in that pan the herrings were baked, fried, or however cooked in."

While some chose to cut off the heads and tails of the fish prior to cooking, the more common practice left them intact (they could be removed prior to serving). "Yes, you are quite right," Edith Austin Holton told amused readers in 1944, "the heads and tails were still there. So were the scales, and all the insides." The fish would then be roasted for upward of half an hour or more, by which time, "and for quite awhile before, everybody on the street would know you were having herring for supper."

Eating Herring

Eating herring was humorously recalled by Edith Austin Holton in her memoirs of growing up in the 1880s and 1890s in Falmouth, Massachusetts:

Each person took a whole fish on his plate, slit the skin along the belly, and expertly peeling it back, lifted out the edible flesh, bones and all, and the red roe if you were lucky enough to draw one. Some people, it is true, preferred the alternate method of eating out of the skin direct...The main trick was not to notice the parts you didn't like the looks of. There was no way of leaving a herring plate tidy when you were through with it.

Frequently, little else was served with a herring meal, except for bread or rolls with butter, followed by pie. Pickled cucumber slices often accompanied the meal. Lorenzo Wood recalled readers telling him of such fare at community herring bakes that also included a "handy" glass of milk "in case a stray bone should disturb the esophagus." Ostensibly the bread was served for the same purpose.

Regardless of how they were prepared, river herring proved a popular staple food locally throughout the nineteenth century. James A. Burgess of Middleborough recalled his nineteenth-century youth and the fact that herring were readily consumed. "People ate them in those days and it was no disgrace." In fact, many found herring an economical meal—always an attraction for frugal New Englanders. "When a large family can make a delicious dinner on 6 [cents] worth of fish, well may the people rejoice and set Brighton [Boston's cattle] market at defiance." Fannie Farmer similarly termed herring "a most economical food."

The great demand for herring as a food staple was cited as early as 1860, when the already appreciable decline in the local herring fishery contributed to its seasonal demand. "The eagerness to take this desirable fish, and the increased number of days in which they can be taken has diminished their number." One later local partisan of herring was *Middleboro Gazette* editor Lorenzo Wood, whose column "By the Clear Nemasket River or Echoes from Shad Row" made frequent mention of not only his great love of the fish but also his impatience for its arrival each spring.

Others had a different take on the fish altogether. The social stigma that eventually came to surround the consumption of herring notwithstanding, some simply found the fish unbearable to eat, either because of its many bones or because of the taste imparted by the means through which it had been preserved. Sarah Orne Jewett, the noted New England author, was critical of both herring and shad for their bony nature, and she opined that they were "about as good to eat as a rain-soaked paper of pins." Joseph C. Lincoln in 1935 related an anecdote whereby his doctor anticipated the annual herring run for the income it would bring him. "Scarcely a day passed, he used to declare, that he was

[not] called upon to extract a herring bone from a child's throat." And while Lincoln suspected that the physician was prone to exaggeration, the tale nonetheless points up the reason for the aversion toward the herring that many held: its excessive bones. The *Brockton Enterprise* once described the fish as "95 percent bones, 5 percent meat." For his part, Lorenzo Wood seemed unperturbed by the bones. "He still declares, with firm conviction, that the bones are no problem."

Another New England author, Sarah P. McLean Greene, undoubtedly recounted her own aversion for the fish in the largely autobiographical novel *Cape Cod Folks* (1881), which drew on her experiences at Plymouth, which she fictionalized as the town of Wellencamp in her writing:

Sarah P.M. Greene, 1881. Not all southeastern Massachusetts residents liked herring. Greene unflatteringly described a herring breakfast in her semiautobiographical *Cape Cod Folks*. *Author's collection.*

> *Slit herrin' was a long-dried, deep-salted edition of the native alewife, a fish in which Wellencamp abounded. They hung in massive tiers from the roofs of the Wellencamp barns. The herrin' was cut open, and without having been submitted to any mollifying process whatever, not one assuaging touch of its native element, was laid flat in the spider, and fried.*
>
> *I saw the Keeler family, from the greatest to the least, partake of this arid and rasping substance unblinkingly, and I partook also. The brine rose to my eyes and coursed its way down my cheeks, and Grandma Keeler said I was "homesick, poor thing!"*

As part of the meal, the protagonist also had golden seal diluted in milk and sugar, but "the herrin' had destroyed my sense of taste; anything in a liquid state was alike delectable to me…. When I got up from the Keelers' breakfast table there was something choking me besides the herrin' and golden seal, and it wasn't homesickness, either."

Later, the protagonist, a teacher, noted another drawback to herring: the brine in which the fish was preserved created a powerful thirst that was shared by her pupils:

> *I was loudly importuned on all sides for water. I was myself extravagantly thirsty. I requested all those who had "slit herrin'" for breakfast to raise their hands.*
>
> *Every hand was raised.*
>
> *I gravely inquired if slit herrin' formed an ordinary or accustomed repast in Wallencamp, and was unanimously assured in the affirmative.*

So distasteful, in fact, did many find the fish that the comment was often made that rather than consume the bony fish, they would prefer to eat the brown paper on which it was traditionally cooked, a suggestion that left Lorenzo Wood aghast.

WEST INDIES HERRING

In addition to the herring that were captured for home consumption, millions were shipped overseas, primarily to the West Indies, where they represented an inexpensive food source. "Years ago, when the herring fisheries were at their best," wrote local newspaperman James Creedon in 1909, "large hauls were the rule, and the fish were dried, smoked or salted for consumption and hogsheads of them were shipped to the west Indies."

Salted or pickled herring became a staple food source in the Caribbean islands during the colonial era. In 1763, Massachusetts shipped some 10,000 barrels of "shad, alewives and other pickled fish" to the West Indies, representing a value of £5,000. The period between April 1, 1804, and January 1805 witnessed the inspection of some 19,163 barrels of pickled fish destined for export from Massachusetts. During the 1830s, the fish were selling for $1.50 to $2.00 per barrel in the islands. Imports into Jamaica in 1849 included 4,220 boxes of alewives and 15,230 boxes of herring. Taunton River herring also were shipped to Bermuda during the late eighteenth century.

In Middleborough, however, the West Indies herring trade seems to have received a later start. In 1764, when the town opted to auction the rights to the herring privileges, it stipulated that "whoever bought the fish privilege should not pickle for the foreign market." Additionally, at the start of the nineteenth century, the numerous and constantly changing restrictions that were enacted by Massachusetts on the pickling and export of fish also dissuaded Middleborough residents from the practice.

Over time, fewer restrictions were set by the town regarding the manner in which herring could be disposed, and many would ultimately be processed for the lucrative Caribbean market. Unsurprisingly, David A. Tucker was foremost among local West Indies herring purveyors. "At one time his exports of smoked and corned herring to the West Indies were very large." Behind the American Hall on South Main Street, Tucker operated a smokehouse in what has been described as "a sizable shed." "There [herring] were processed for their long trip to the island market. One of the prize jobs for youngsters who were lucky enough to get them was 'stringing' herring on sticks."

The West Indian market would eventually evaporate as other food sources were found. Although Nemasket herring were once more shipped to the islands in 1910, the shipment by then was considered so unusual as to warrant an item in the local press.

The Social Stigma of Herring

In time, an aversion grew among New Englanders toward herring, which became known as "poor man's salmon," a characterization seemingly supported by the fact that several thousands of the fish were consumed annually at the Middleborough poor farm. Additionally, the practice of annually granting "poor widows" two hundred of the fish free further reinforced the connection between herring and the community's less fortunate. Even when the fish were available free, some were "too proud to carry away a string of them through the streets," so closely had the fish become associated with poverty. For those who could afford to do so, shad replaced herring.

A number of authors of the era reflected the increasing stigmatization of herring in their works. In *A New England Nun*, one of Mary Eleanor Wilkins Freeman's characters responds, "I don't want no herrin's, now we've got this honey," but she chooses to eat the herring, which she secretly relishes, on the pretense of not wanting to waste money. Similarly, the prolific output of regional novelist Joseph C. Lincoln also reflected the growing association between herring and poverty. One character in Lincoln's 1916 novel *Mary-'Gusta*, Solon Black, recounts the tale of his voyage on a small sailing vessel in the company of Captain Jerry Clifford, who owns property and so "can't be a poor man": "Solon swears that all the hearty provision Jerry put on board for a four-day trip was two sticks of smoked herrin'. For two days, so Solon vows, they ate the herrin' and the other two they chewed the sticks. That may be stretchin' it a mite, but anyhow it goes to show that Jerry Clifford don't shed money same as a cat does its hair." Clearly, the consumption of herring had moved from an act of necessity to one of frugality.

Thanks to such literary works, the fish came to be increasingly associated outside New England with the stereotype of the parsimonious, parochial and peculiar Yankee—an image most Middleborough residents sought to avoid and one that was increasingly giving way in consequence of the flood of immigrants flowing into the region. With the increasing availability of other fish, fewer and fewer residents exercised their option to take herring. Yet ironically, as many strove to distance themselves from the eating of the fish and the negative connotations of doing so, others continued to cling to the old ways not only because they favored the fish but also because the practice was a stamp of their unique identity as New Englanders, both thrifty and hardy. Among them was the Wood family, who for many years owned and edited the local *Middleboro Gazette*. In 1910, the *Gazette* impatiently awaited the herring's arrival: "A little more warm weather should soon bring them up as far as Middleboro, and then ho, for a feast!" For a minority of the newspaper's like-minded though slightly more dishonest readers, the spring run provided the opportunity for a free meal. So eagerly did they anticipate the return of herring in the spring that they were sometimes negligent in complying with the laws regarding poaching. In April 1910, "unbeknown to the fish wardens, many of the residents enjoyed this toothsome dish cooked in real old New England style the first of the week."

HERRING AND MIDDLEBOROUGH SOCIETY

Despite the social stigma that developed around the consumption of herring locally during the nineteenth century, it was during this period that the fish came to be wedded with the community's cultural consciousness.

Herring fishermen in the seining pool below the Star Mill dam, 1950s. *Middleborough Public Library.*

The familiar cry of "Herring have come!" became ubiquitous each spring, being heard in the streets, in stores and churches and anywhere else residents gathered. The phrase received equal play on the front pages of the local newspapers, including both the *Middleboro Gazette* and the *Middleboro News*, both of which gave full notice of the fish's annual arrival. Then, as now, the herring were regarded as harbingers of spring: "Not the sight of the early robin or the fluffy pussy willow or the goat rampant on the lithographed wall sets us a tingling with the pulses of spring as does the cheerful cry, 'Herring's come.'"

The cry, "Herring have come!" was a phrase bound to catch any Middleborough resident's attention and was quickly adopted as an advertising catchphrase, though frequently with little connection to the merchandise or service advertised. Local fish dealers, however, found it useful when advertising their own catch. In early April, at the start of that year's run, H.H. Adams, proprietor of the Middleboro Fish Market, advertised, "Herring Have Come! Native Herring. Fish every day. Shad, Fresh Tongue, Spawn."

Herring were also responsible for another popular regional phrase of the nineteenth century—"Taunton, Good Lord!"—heard in Middleborough and spoken of its neighbor to the west. Some attribute the phrase to sailing captains bound for Taunton immediately prior to the annual herring run who frequently answered the question of where they were bound with the reply, "Taunton, Good Lord!" in anticipation of the fresh herring on which they would soon be feasting.

The arrival of the herring not only signaled the end of winter but also brought to an end what seemed like months of interminable waiting, especially for those particular to the fish, who welcomed not only the ready abundance of herring but also their consequently low prices. As early as March 1747, almanac writer and publisher Nathaniel Ames of Dedham had written, "[A]nd now out the frogs peep which forebodes that alewives will be cheap." Just 130 years later, although the first day's catch in mid-April 1877 yielded only four hundred of the fish, the fact that herring had finally come prompted the *Gazette* to almost sigh, "[W]e are bound to live now even if we do not lay up a cent." The syndicated tale "Roweny in Boston," published in 1891, explicitly associated herring and spring, noting that "sure sign that spring was well advanced, men were coming up from Taunton with new herring for sale; it was now so late that the herring would be cheap enough to buy." In a similar vein, Lorenzo Wood in 1959 would write, "Our spirits have been rising of late. Frequent banquets of native herring, the song of the peepers in the air, blended with the perfume of the skunks' cabbage really rids one of the wintertime blues."

With the start of the annual herring run, eagerly awaited by many, it is not surprising that the arrival of the fish was heralded in local newspapers, which published accounts of the first sightings. Tellingly, during the mid-nineteenth century, when few local news items were headlined in the local *Namasket Gazette*, the arrival of the herring in 1855 was headed "The First Herring."

Herring remained front-page news for the next century, as newspapermen followed their comings and goings and (increasingly with alarm) their dwindling numbers. The arrival of the fish in great numbers was frequently depicted in a pious age as a gift of Providence. The religious editor of the *Middleboro News* commented exuberantly on the run in mid-May 1884:

> *There was an unusually large catch of herrings at the Star Mills on Monday, there being about eight thousand caught. This was the first day that the catch has amounted to anything large. Verse 57.—"Thanks be to God." A most natural outburst of grateful joy in view of this glorious victory.*

Such sentiments harkened back to the days of the colonial fishery and the years of Middleborough's first settlement, when undoubtedly such thanks *had* been given for the Nemasket's bounty of fish.

Star Mill harvest. Photograph by George Morse, 1910. Fishermen have closed the top of the fishway and are harvesting herring as they back up. *Author's collection.*

Once the herring began to run, it became an all-consuming preoccupation with the community. One local folktale relates an episode in which a Sunday sermon was interrupted by a woman, bursting into the church carving knife in hand, who announced to the startled congregation that the herring had come, whereupon the remainder of the sermon was tabled and the congregants (presumably accompanied by their minister) made a mad dash to the river. Such springtime passion for the fish was prevalent throughout the region. "Would you arouse the enthusiasm of a shop-weary Brocktonian, whisper in his ear some bright April morning that…the Herring are up," advised Pembroke historian Henry Wheatland Litchfield in 1909.

Not only the arrival of the herring but also their immediate preparation was a further indicator that spring truly had arrived. In March 1898, the *Middleboro Gazette* reported that "the pungent and wholesome odor of the spring bonfire and the roasting alewife now scent the air." Edith Austin Holton similarly recalled as a sign of spring "the smell of roasting herring, pungent, almost acrid, potent enough to revive the most jaded appetite."

The first meals of the season of fresh herring were anticipated by many. Even Thoreau was excited at the prospect of a meal of Taunton River herring. In a letter dated March 17, 1838, to his brother, John, who was teaching at Taunton, Thoreau wrote in acknowledgement of a pacakge received, "Dear John—Your box of relics came safe to hand, but was speedily deposited on the carpet, I assure you. What could it be? Some declared it must be Taunton herrings: 'Just nose it, sir!' So down we went onto our knees, and commenced smelling in good earnest—now horizontally from this corner to that, now perpendiculalry from the carpet up, now diagonally—and finally with a sweeping movement describing the circumference. But it availed not. Taunton herring would not be smelled." It ended up being a box full of another noted Taunton product: nails.

Another story is told of a student at Brown University, a favorite school for Middleborough scholars, who sought (and were denied) permission to travel home for Thanksgiving. Following the refusal, he requested that he have a herring for dinner, whereupon the now enlightened master responded, "Oh, if you wish to go home for a herring, you shall go. So important a request cannot be denied."

In what developed as a tradition, local residents frequently brought the first herring of the season to the editors of the local newspapers. During the second week of April 1861, George S. Savery provided the staff of the *Middleboro Gazette* with "a string of nice fresh herrings, part of the first catch of the season." Later, in mid-March 1885, editor Sylvester of the *Middleboro News*, "who is always having something good happen to him…had the first taste of fresh herrings this season." The Woods of the *Gazette* were extremely fond of herring. In 1913, fish dealer J. Herbert Cushing remembered "the publisher's weakness, and…brought in the first herring of the season."

While the annual herring run was eagerly awaited by local gourmands, most grew tired of the fish quite quickly:

> *Not that we stayed lyric over our herring for long. By the time the twilights had reeked of them for a month or more the most confirmed addict responded somewhat sluggishly to the supper call. This does not include the habitual herring eaters, of course, but the rest of us were pretty blasé.*

Despite the fact that some did not favor the fish for eating, the herring was held in high regard in the community, and quick umbrage was taken at any slur—explicit or implied—as illustrated by an anecdote presumably dating from nineteenth-century Middleborough and retold by Lorenzo Wood of a noted character:

> *He was a brilliant man, though dissipated, and upon his return from a term in Taunton jail for drunkenness was asked the reason for his incarceration and quickly replied: "Well in a discussion, I made the remark that there were as good fish as herring, upon which they put me in jail."*

The sight of herrings strung on a stick eventually became an iconic image of New England, one that novelist and Cape Cod colorist Joseph C. Lincoln depicted in his novels and recalled more particularly in 1935:

> *The sticks were for the most part whittled from cedar—we are quite sure cedar was the wood used. Made from the old split-cedar fences that used to be so common…. The herring sticks were pointed at both ends and whittled thin enough to pass through the gills of the fish. A dozen were strung on a stick.*

The local practice called for the more gruesome method of passing the stick through the fishes' eye sockets, as recalled by Lorenzo Wood:

> *Apparently preserved for all time by the smoking process, the stiff and yellowed forms of our finny friends from the ocean depths hung from long stout sticks whose pointed end had pierced the eyes of the herring, assuring a secure fastening of the coming meals.*

Regardless of the method employed, the fish would later be hung in barns, outbuildings and sheds high in the rafters, where they were a common sight throughout the region, as noted by both Wood and Lincoln. In *Cy Whittaker's Place* (1908), Lincoln depicts "the vegetable cellar where the sticks of smoked herring hang in rows above the barrels of cabbages, potatoes, and turnips."

The stringing of herring became a common cultural phenomenon, one known to nearly all New Englanders, and it was custom by which they might be identified. One anecdote from the Civil War concerns a Massachusetts soldier who fired his gun, having neglected to remove the ramrod first. "His comrades frequently joked him afterwards, in regard to this incident, facetiously saying, that he was so accustomed to stringing alewives and other fish on rods, that he was trying to string rebels on his ramrod."

The image of herring strung on a stick would become widespread with the introduction of picture postcards and would appear in books on New England. Not surprisingly, one such view is depicted as the chapter head in Weston's own *History of Middleboro*.

Herring attracted devotees and staunch protectors. Foremost among Middleborough residents were the irascible James A. Burgess, who had a soft spot for animals of all kinds, and James A. Thomas. In nearby Mattapoisett, Alexander B. Bowman could be found, conspicuous in his white derby, observing the herring runs during the last twenty years of his

James A. Thomas, circa
1890. Of Thomas,
Lorenzo Wood once
asked, "Was there ever
a more ardent supporter
of the protection of the
[herring] than Jim?"
Author's collection.

life, while Patrick Gibney visited the East Taunton run every year following
his immigration from Ireland in the 1860s.

The humble herring not only marked the New Englander but also tended
to separate the rural Yankee from his more cosmopolitan counterpart. One
verse attributed to New England extolled the simplicity of the local fish over
the more urbane seafood dishes that were favored in Boston:

> *When I went up to Boston*
> *They served me fish with frostin'*
> *But heavens it was costin'*
> *Too much for the likes of me.*
> *Give me herring and potatoes*
> *Herring and potatoes*
> *Herring and potatoes*
> *Is enough for the likes of me*

Herring were also regarded as increasingly a country phenomenon unknown to those of the city:

> *A gent from the city whom we had come to believe might someday become "one of us" and enjoy the better things of life, simply turned up his nose [at herring], said the whole mess stunk, and refused to try that which he might well have added to the brain power which he will need in the future.*

To those who sneered at those two most important of Massachusetts fishes—the cod and the herring—locals might respond:

> *What we eat until we're bustin'*
> *Manitobans find disgustin'*
> *For it seems salt cod's upsetting*
> *To their dainty western tongues*
> *And with faces stern and stoney*
> *They say herring's too damned boney*
> *But there's nothing for dissolving bones*
> *Like good Jamaica rum.*

Herring entered the Middleborough vernacular and appeared in a number of uniquely Yankee metaphors, being used to describe everything from old age, thinness and dry personalities to death. Joseph C. Lincoln, whose novels came to epitomize conceptions of New England, and Cape Cod in particular, employed a number of these metaphors, as did several other New England regional authors, thereby reinforcing the connection between coastal New England and river herring.

These same regional works also reinforced the historical continuity of Middleborough or Nemasket's role as the "place of fish." Albert Hale Plumb wrote romantically of Middleborough's colonial days in his *When Mayflowers Blossom: A Romance of Plymouth's First Years* (1914): "Outside the vale of Nemasket, that is, the place for catching fishes." References to herring became increasingly apparent in such works, ironically at a time when the numbers of the fish were dwindling and runs outside Nemasket were dying. As the number of herring continued to decrease each successive season, the community's attachment to the fish as a cultural icon became stronger and its identity more closely wedded to it.

THE DECLINING FISHERY, 1901–1920

THE NEMASKET FISHERY IN DECLINE

As early as 1860, the *Middleboro Gazette and Old Colony Advertiser* noted a drop in the annual herring run and attributed it to "the eagerness to take this desirable fish, and the increased number of days in which they can be taken." Throughout the late nineteenth century, overfishing of the Nemasket and Taunton Rivers, a consequence of irresponsible management, was blamed for compromising the local fishery and faulted for the extirpation of the American shad from the Taunton River watershed. Although legislation had existed in Massachusetts since 1804 to protect mackerel from overfishing, no such protection was afforded the herring, and laws relating to the latter fish focused nearly exclusively on securing adequate upstream passage during their annual spawning run. Efforts to establish a salmon fishery during the 1870s in the Taunton and Nemasket Rivers, supported by the commonwealth, may have been a tacit acknowledgement that the declining herring fishery was in trouble.

Some (like Thoreau, who was concerned about herring being "driven out of a river by the improvements of the civilized man") regarded the fish's declining numbers in spiritual terms. "No doubt there is some compensation for this loss," Thoreau thought, "but I do not at this moment see clearly what it is…It is as if some vital quality were to be lost out of a man's blood and it were to circulate more lifelessly through his veins." Others, among whom were many Middleborough residents, were more troubled by the economic implications of decline and its impact on municipal employment, tax mitigation and poor relief. In 1883, the herring decline was placed in context for the townspeople: "The alewife fishery which in 1867 was an income to Middleboro' of $1552, has since then gradually declined year by year, until last season [1882] the fishery sold for but $220." Additional economic concerns were cited, including the fact that herring "furnish[ed] a large part of the inhabitants with cheap and wholesome food

Henry David Thoreau, 1856. Thoreau saw the destruction of herring runs as man-made and sensed their loss deeply. *Author's collection.*

in its season, and [gave] employment and revenue to quite a number of persons, engage[ed] in the taking and sale of the fish."

In the years immediately following 1900, the decline in the local herring run garnered increasingly more attention. In 1903, the *Taunton Gazette* noted that "while some of the Taunton river fishermen complain of a scarcity of herring it is apparent that the fish go up the river and dodge the seines in some way, or they would not reach Nemasket and make it a valuable fishing place." Yet the herring, no matter how wily they may have been in eluding nets along the Taunton River, were not as populous in Middleborough as Tauntonians believed. James Creedon, writing for the *Brockton Times* in April 1905, attributed declining receipts of the Nemasket fishery to the fact that "the number of herring which come up the river is not as many as formerly." Thomas Weston, in his authoritative *History of the Town of Middleboro*, published the following year, was likewise cognizant of the decline, as well as the possibly dire future of the Nemasket fishery: "Of late the herring have from different causes so decreased in number that the amount received by the town is small…. The last few years would indicate that the time is not far distant when the herring of the Nemasket River may become so far extinct as to cease to provoke much attention and action on the part of the town."

Factors for the decline other than overfishing were considered as well. In 1907, the *Middleboro Gazette* suggested that the Wareham Street dam was to blame, its dilapidated condition not permitting passage of adequate numbers of fish. More unusual suggestions were also put forth. In 1912, despite heavy catches at East Taunton, those at the Star Mill privilege remained light, a circumstance then "attributed to the engine on the railroad bridge at Muttock, which had a tendency to retard the progress of the fish." Presumably the fish were deterred by the communication of vibrations from the locomotive to the water below.

The decline in the Nemasket fishery became critical during the second decade of the twentieth century. The *Middleboro Gazette* at that time characterized the situation as alarming and noted that it had reached a climax in 1916, "when the largest part of the Taunton river fishing privileges [including those at Middleborough] were not operated, even though the market price for the herring was the highest in years."

Reports of the dwindling numbers of herring only exacerbated the financial precariousness of the local fishery, and Middleborough selectmen struggled to attract a lessee for the town's privilege, so widespread had knowledge of the fishery's decline become. In 1915, selectmen received only one bid for the fishery, from Erving B. Merrihew of Wareham, with whom they subsequently had difficulties as he refused to pay anything beyond the initial ten-dollar down payment. The following year, again, only one bid was received, "and this was so low that they…decided not to dispose of the right [that] season, giving the fish an opportunity to have a clear way to the lakes that they may be more numerous in the future"—undoubtedly soured on the previous year's experience.

Continued light catches in 1918 were attributed, in part, to wartime measures, including the emplacement of antisubmarine and mine nets in Narragansett Bay and the testing of depth bombs there as well, the latter practice "known to have killed millions of herring." Increasingly, however, some observers believed that a more insidious factor was at work.

By 1920, the state of the Taunton and Nemasket River herring fisheries was stark. In that year, under a headline that read, "Taunton Is Herringless," the sad condition of the Taunton River was noted. "Taunton river, which a few years ago had a herring and shad fishing industry more extensive and important than all of the others combined, is a dead one for supplying fish and the recent seasons at Dighton and Somerset have been absolute failures." Dighton, in fact, gave up its fishery in 1925 after receiving just one dollar for the privilege in both that year and the preceding year, a drastic change from the days when it could be claimed that the herring fishery had brought more wealth to Dighton than any other industry. Similarly, the other downriver communities abandoned their fisheries during the same period for lack of fish. For Middleborough, such developments were truly alarming.

A Report Upon the Alewife Fisheries

In 1920, in response to declining populations statewide, biologist David Belding conducted a landmark survey of Massachusetts' herring fisheries for the state Department of Conservation. Entitled *A Report Upon the Alewife Fisheries of Massachusetts*, its conclusions were not encouraging. According to Belding, many herring runs in the commonwealth had been completely destroyed and were beyond reclamation, while others were either greatly impaired or operated below their full potential.

Of the Nemasket fishery, Belding wrote that it had been compromised "undoubtedly due to the new uses to which the river was put as a means by which to carry away the town's sewage and manufacturing effluent. Additionally, developments in the Taunton River both above and below the Nemasket over which Middleborough had little or no control impacted the Nemasket fishery negatively."

Contrary to the view that overfishing was dooming the Nemasket fishery, Belding cited three sources for the decline: "Neglect in keeping fishways in proper shape, permitting pollution such as sewage and manufacturing wastes to enter the stream, and the illogical method of leasing the fishery for a one year period."

NEGLECTED FISHWAYS: WAREHAM STREET

Partly in line with Belding's conclusions was the view held by local officials that inadequate fishways at East Taunton and Wareham Street in Middleborough were the principal cause of the Nemasket's decline. "The improvement of the [Wareham]-st fishway, and also the one at East Taunton, both of which are said to be difficult for the fish to pass through…will remedy the difficulty," wrote a hopeful correspondent in April 1905.

For decades, breaching the Wareham Street dam in spring had been a challenge for the herring, and much blame was placed on the 1883 Brackett fishway. In 1895, "another and even better channel [was] provided at the old sawmill waterway" for the fish, and this was described two years later by the *New York Times*:

> *Owing to the great declivity of the ground, an ingenious waterway has been constructed, which allows the fish to pass by easy stages from below the falls to the mill pond above. A trough has been made about 150 feet in length, 3 feet broad, and 3 feet deep, and placed at an angle of 30 degrees, connecting the mill pond with the water below the falls. The water rushing through here comes with such force that the herring would have to make too long and sustained effort in order to swim through; so planks extending half way across the trough are placed alternately, thus making a series of shelves behind which the fish find resting places.*

Nonetheless, the 1895 ladder was not much of an improvement, and changes were made to it by the 1900 lessee, Charles Noble Simmons of Dighton. These failed as well, and Simmons's employees were compelled to transport some fifteen thousand herring over the dam from the base of the weir to the millpond on the opposite side. Three years later, in 1903, the fish still faced problems on their upstream run:

> *The alewives have encountered their usual tribulations in journeying upstream to the lakes. There has been no water in the boxes at the [Wareham] street fishway, owing to the fact that the head of the dam was broken in, so the progress of the fish has been retarded and almost stopped.*

H.L. Leonard was subsequently engaged to put the fish over the dam.

Wareham Street dam, circa 1910. The dam long posed a challenge to the herring in their migration until it was finally rebuilt in 1964–65. *Author's collection.*

In the spring of 1907, at a time when the Middleboro Business Men's Club acknowledged that the Nemasket herring run, "one of the best and most profitable herring runs of the state," had "so deteriorated...as to become practically worthless," herring again had to be transferred to the upstream side of Wareham Street. Although the club's viewpoint that herring had "rights...to pass up Nemasket river" that might supersede those of men was unprecedented, its call to improve the Wareham Street site fell on deaf ears; some forty thousand herring had to be seined from the river below the dam, "carried across the dam in tubs and dumped on the other side." This method, however, still left large numbers of fish stranded below the dam, with disastrous consequences:

> *At the present time the water around the fishway, which is not in running order, is black with herring which cannot go further owing to not being able to climb the ladder. Around this fishery there are large numbers of dead herring, lying about on the rocks and banks of the river, and these give out a very disagreeable odor. The fish have been disabled by being dashed against rocks in the swiftly running water or against the posts which hold up the fishway, which is practically useless, or having been caught by boys, only to be thrown upon the banks and left to die.*

Only after the *Middleboro Gazette* chided the Massachusetts Fish and Game Commission and a threat of legal action was made did work finally commence on construction of a new fishway in the spring of 1908. The concrete ladder constructed by local contractor Martin O. Rounseville consisted of "four one foot jumps for the herring, six feet apart." A later account was more precise, describing the ladder as

> a series of cement boxes of about six feet wide and 12 feet long, and when full of water from four to five feet deep. Curved depressions in the mid-section of each, provide an opening for the fish to dodge through when water is running in the stream. The total length of the fish ladder is about 300 feet.

The optimism surrounding the new fishway quickly evaporated when it was discovered that the fish were unable to pass through it. "They are still on the same old job, that is fixing the fishway, and they don't get past even the new one," noted a despondent James Burgess at the time.

Temporary expedients, such as sandbagging the river and the emplacement of a concrete cofferdam in 1909, were succeeded in 1918 by construction of a new, successful ladder designed by Middleborough fish warden Daniel F. Wilbur:

> The new section is of concrete construction and the upper wall was built about three feet higher. Steps were constructed south of the bridge on the same lines as those on the other side and these give the fish plenty of opportunity for resting. By the new wall higher water is maintained in the river and the municipal light plant can have the full advantage of the water power.

NEGLECTED FISHWAYS: EAST TAUNTON

Middleborough authorities also believed that the Nemasket herring population was being adversely affected by the Brackett fishway at East Taunton, described by former Tauntonian Alfred J. Pairpoint in 1891:

> To assist the herring tribe, the renters of the waters have invented what is called "The Boxes"; a sort of plank division, of strong woodwork, is built and fixed securely in a kind of canal rapid, the other side of the waterfall,

thus forming a breakwater in sections, and the velocity of the rushing stream hurls the fish into these safety boxes, where the finny creatures rest preparatory to another venture into the next division of boxes, where the fish pant and recover their strength for other attempts to go on for a few yards more, when the brave fellows get into the smooth water of Taunton River.

Herring had difficulties in climbing the ladder, and in addition to its deficiencies in providing upstream passage for the herring, the East Taunton ladder with its patented box arrangement also proved inadequate for shad, which showed a decided disinclination to ascend it.

As early as 1905, Middleborough was calling for a new ladder at East Taunton. Demands intensified after 1910, and in 1915, James A. Thomas urged that the East Taunton fishway be "put in proper condition for the shad to go up, which many of the older people in the vicinity know to be feasible, notwithstanding the wisdom of the fish and game commission."

In 1917, the East Taunton fishway finally was replaced with a wooden fishway designed by Deputy Fish and Game Commissioner Allen David:

With its sloping bottom and irregular baffles it resembles the Brackett type, but possesses the additional qualifications of frequent rest pockets and a steady, uniform flow of water which is controlled by the upper gate. Although more expensive than the second standard type, it can be advantageously installed in a limited space over an irregular course. [The new fishway at East Taunton was designed to] *climb over a 10-foot water head during a hundred foot lateral progress, with a 10 per cent grade. The circuitous route of the fish carries it 325 feet in advancing a hundred feet.*

The additional problem of juvenile herring being drawn into the penstocks of the Connecticut Cotton Company that stood alongside the East Taunton dam was addressed at this time as well, by forcing herring fry to return through the ladder. "Millions it is claimed, are ground up by the powerful turbine wheels of the cotton mill, through which, in August, September and October practically all the water passes." Screens were installed thereby directing down-running fish "through the same boxes which the alewives pass through on their way up earlier in the season."

Much optimism surrounded the development of this new ladder style, designated the David fishway, which was expected to materially improve the herring fishery "and bring it back to its old time volume." So enthusiastic

Brackett fishladder, East Taunton, circa 1910. Herring required eight hours to ascend the Brackett ladder. They could climb its replacement in just fifty-nine minutes. *Author's collection.*

David fishladder, East Taunton, 1926. James A. Thomas was critical of this 1917 ladder, characterizing it as "money thrown away." *Author's collection.*

was the state Fish and Game Commission that even prior to the completion of the East Taunton ladder, it proposed the installation of David fishways throughout the commonwealth, including one on the Nemasket at Wareham Street. (It was never built.) The East Taunton David fishway served its purpose for twelve years. In 1930, the East Taunton dam collapsed, thereby removing any need for a fishway.

Neglected Fishways: Star Mill

In contrast to the fishladders at Wareham Street and East Taunton, the Star Mill fishway was well maintained and did not impede the passage of the herring. At the time the Star Mill site was acquired by the Nemasket Woolen Company in 1906, it was described as including "a stream which at one time was the course of the fish upstream, and was also the site of the fishing place." In 1920, the fishway was described more fully as being "in the form

Star Mill dam, circa 1940. The wooden dam stood thirteen feet above the level of the river and featured a natural fishway. *Author's collection.*

of a natural stream of a gradual rise, equipped with stone projections to enable the alewives to pass up against the current."

Threats to herring at the Star Mill site stemmed not from the physical condition of the fishladder but rather from the presence of lawbreakers. In 1907, Walter D. Blair, a Middleborough fish warden as well as that year's lessee, encountered difficulties in protecting the fishery from poachers. On May 5, a Sunday afternoon, "about 15 boys were busy catching herring at the Star mill run where the water was black with these fish as far as the eye could see, when Inspector Blair made his appearance. The boys scattered in all directions, but one, however, was caught and given to understand that he would be prosecuted if found catching herring again. Up to the present time no one has been brought into court charged with this misdemeanor."

The subsequent year, during the 1908 run, it was reported that "a close watch is being kept for poachers, as in years past there has been considerable fish stealing from the weirs." The annual watches appear to have been a wise investment, for in 1910 Warden D.F. Wilbur reported that "the herring stealing has been kept to its lowest this year, and the wholesale night seining of fish and carting them away in team loads has not occurred here." Nonetheless, poaching and harassment of herring remained the primary concerns at the Star Mill site for many years.

Assawompsett and Quitticus

A new threat to Nemasket herring was the acquisition of water rights in the Assawompsett Pond Complex by the cities of Taunton and New Bedford, both of which began to draw on the ponds as municipal water supplies in the late 1890s. Twenty years earlier, the Massachusetts legislature had authorized Taunton to draw water from Lake Assawompsett and permitted the city to erect a dam at the head of the Nemasket River two and a half feet above the existing mudsill to control outflow from Assawompsett, with the proviso "that the natural flow of said Assawompsett Pond into the Nemasket River shall at all times be maintained." No action was taken to build the dam, however, until 1894.

Concurrent with this action, New Bedford also began to look to the Middleborough ponds as a public water supply for the city. Following 1870, the failure to keep sufficient water in New Bedford's reservoir at Acushnet

Assawompsett dam, circa 1920. The 1894 dam obstructed herring and affected water levels in the Nemasket River, with sometimes disastrous consequences for the fish. *Author's collection.*

had proved problematic, and by 1886, water levels there had fallen so significantly as to permit the exposure of vegetative matter to the sunlight, creating foul-tasting water. To remedy the situation, a channel was cut linking the reservoir with Little Quitticus Pond in nearby Rochester. "New Bedford, to replenish its failing water supply, has tapped Little Quittacus Pond and now the question arises concerning damage to the various...privileges on Nemasket River." This arrangement, designed only as a temporary expedient, existed until July 1899. Although proposals called for upgrading the Acushnet reservoir, that facility was considered inadequate to meet the needs of the growing city; consequently, a completely new waterworks system for New Bedford, utilizing Little Quitticus Pond as a terminal reservoir, was constructed in 1899 from plans drawn three years earlier.

When New Bedford and Taunton's manipulation of the water levels of the Middleborough ponds resulted in a reduced volume of water in the Nemasket River that hampered both the downstream migration of juvenile herring from Lake Assawompsett as well as the operation of Middleborough's Municipal Light Plant, Middleborough brought suit against the Bristol County cities in 1903. Following six years of litigation, Middleborough won its suit, although the cities' rights in the ponds were preserved, much to the detriment of the herring.

Long Point Road, circa 1910. In 1899, New Bedford rebuilt the causeway to prevent outflow from Great Quitticus (right) into Pocksha Pond (left). *Author's collection.*

In 1907, it was noted of Great Quitticus that "formerly large quantities of alewives went into it through a small brook from Assawamsett Pond, but very few pass now." Belding in 1920 explained the reason for Little Quitticus's low proportion of herring: "The constant use of the water…reduces it to such an extent that during the period of low water in the fall the young alewives are unable to leave the pond, as at Great Quitticas Pond." As a result of not being able to leave Little Quitticus, juvenile alewives clogged the city's intakes and were killed by the millions. It was recalled in 1916 that "one day a year or so ago employees of the New Bedford water works removed over 50 barrels of the young fish that had died." To avoid this, both Great and Little Quitticus were sealed off from the fish as spawning grounds in 1916, when a screen was constructed between Pocksha and Great Quitticus Pond "of such a size as to prevent the full grown fish from getting through." Fish were no longer able to pass into Quitticus to spawn.

"DEAD ALEWIVES IN ABUNDANCE"

Besides being attributable to inadequate fishways, the late nineteenth- and early twentieth-century decline in the Nemasket herring population was also due to pollution in the form of sewage and manufacturing discharges into both the Nemasket and Taunton Rivers. Beginning with the earliest mills, pollutants in the form of sawdust, wood shavings, grain chaff, textile fibers, wool waste, grease, dyes, wastewater and other items were dumped into the region's rivers.

At the time, the polluted condition of the Nemasket and Taunton Rivers and the potential impact of that pollution on the marine life in those waterways was not well understood. Yet while the precise scientific process by which pollution acted negatively on living organisms was unknown to the late Victorians, the actual consequences were not. As early as 1871, poet James Russell Lowell recognized the connection between pollution and dead fish, writing in "My Garden Acquaintance" of a crow he noticed in his garden: "I do not believe, however, that he robbed any nests hereabouts, for the refuse of the gas-works, which, in our free-and-easy community, is allowed to poison the river, supplied him with dead alewives in abundance."

Pollution of the Taunton and Nemasket Rivers did not go entirely unrecognized during the late nineteenth century. In April 1875, when a spring freshet filled it with raging floodwater, the Taunton River's principal Taunton tributary, the Mill River, was described as "a dirty, foaming torrent of tremendous proportions," a result of that river's heavy industrial development over the previous centuries. The Taunton Board of Health, recognizing that city sewage, drainage from the gasworks and effluent and waste from nonmunicipal sources continued to be discharged into the Mill River, essentially acknowledged it as an open sewer, labelling the waterway "offensive" in 1895 and noting that it had "long been a source of filth."

When in April 1877 the Massachusetts legislature passed an act making it a punishable offense to dump "any sawdust…or any drugs, dye-stuffs, acids, alkalies or any other substance destructive of the life of shad or alewives" into the Weweantic River, which formed, in part, the eastern boundary of Middleborough, it seemingly recognized the detrimental effect these substances had on marine life. Yet the law applied only to the Weweantic River partly because other rivers, such as the Mill River, were considered too far gone in regard to fish life and partly because the Weweantic ran entirely through an unurbanized region where there were no manufacturing interests to object to the ban.

James Russell Lowell, circa 1860. As early as 1871, the poet recognized the connection between polluted rivers and decimated herring runs. *Library of Congress.*

Despite the example of the Weweantic River, any concern that existed for the polluted condition of waterways in the Taunton River watershed was less focused on the environmental impact of pollution and more on the potential contamination of municipal drinking water sources. Even this latter threat, however, did not always prompt communities to act. In 1885, the Plymouth *Old Colony Memorial* cynically suggested that pecuniary interests may have derailed public health measures, indicating that increasing pollution in the Taunton River had value as a rationale for Taunton's acquisition of a new and pure water supply at Lake Assawompsett:

> *Taunton isn't particularly averse to having its river polluted by Brockton sewage, for the Gazette of the former city says there would be a chance to make enough out of Brockton, to pay for bringing a water supply for Taunton from the great ponds at Middleboro.*

For Tauntonians, however, Middleborough's sewer discharges were another matter, and during the late 1800s, Taunton began "to show signs of fear lest its water supply should be contaminated by Middleboro's sewage." Brockton's 1892 proposal to utilize the Taunton River for the dumping of sewage (and with it the likelihood that Taunton finally would resort to Assawompsett as a municipal water supply) placed Middleborough in a quandary. "Whether it is best to help Taunton against Brockton and thereby get [Brockton] into a position to howl against Middleboro's using Nemasket for sewage, or whether to help Brockton drive Taunton into taking city water from the Middleboro and Lakeville ponds, and thereby lower the level of the Nemasket permanently—that is the question, and [Middleborough] doesn't exactly know what to do." The matter was resolved in 1893 when Brockton constructed a sewage disposal system that became a world model for inland communities. Pumped by means of a station located at Campello, the city's sewage was filtered at the edge of the city, thereby eliminating the need to discharge it directly into the Taunton River, as had been proposed.

By the dawn of the twentieth century, attention was beginning to be called toward the impact the polluted Taunton system was having on herring. In 1902, local historian John M. Deane of Assonet recognized the damage done to fish in the Assonet River, a tributary of the Taunton, by pollution. Deane recalled the abundance of fish there "before the waters of our beautiful river were contaminated by the refuse of the Copper Works on Taunton river, and the Bleacheries and saw mills on its own banks," the pollution keeping "most of the fish out of the river." The following year witnessed an increase in the

Campello sewage pumping station, Brockton, Massachusetts, circa 1895. Completion of the plant in 1893 greatly reduced the potential threat to Nemasket herring. *Author's collection.*

East Taunton fishway, circa 1905. At the time, the East Taunton fishladder (left), rather than pollution, was cited for the Nemasket herring's decline. *Author's collection.*

number of dead herring reported in the Nemasket and Taunton Rivers. The *Middleboro Gazette* attributed the fact that the dead fish were "more plentiful than usual" to the force of the flow passing through the East Taunton fishway. "Most of them are hurt by attempting to climb the dam at East Taunton where they are hurled back and beaten upon the rocks. Some of the feeble fish manage afterward to work through the fishway and die between Taunton and Middleboro." While 1903 was, in fact, a year of high water that tended to inhibit the upstream run, part of the die off, in truth, was ascribable to increasing pollution in the Taunton River and its tributaries.

Although some were beginning to awaken to the negative impact of pollution on local herring, efforts to explain the decline remained focused on the inadequacy of existing fishladders, with few people calculating the irreparable damage being done to local rivers by manufacturers, municipalities and institutions located along the Nemasket, the Taunton and other rivers that discharged pollutants directly into the water. As late as 1911, it still was being stubbornly maintained that the box-like fishladder at East Taunton was the singular impediment to restoring "the old time plentitude of herring" in the Nemasket. Consequently, efforts to revive the herring fishery during this period tended to focus on the reconstruction or replacement of existing fishladders rather than stopping and reversing the flow of pollution. In 1923, James A. Thomas, who had long been associated with the Nemasket herring runs, continued to blame "these new fangled fishways."

Not until 1943 was pollution of the Taunton River seriously considered, when Middleborough selectman William G.L. Jacob in October of that year raised questions about the negative impact of pollution on the herring population. Jacob's trenchant questioning must be credited with drawing local attention to the plight of herring and pollution of local waterways that long predated 1943. Yet while the worsening environmental condition of the Taunton River was increasingly a source of concern for Middleborough, the town was generally less inclined to consider how it treated its own river.

Historically, the Nemasket River was fouled by a number of sources, both industrial and municipal. The earliest polluters of the river were colonial mills located along its course. Sawmills dumped sawdust into the river, while the fulling mill at the Lower Factory discharged lanolin-laden fuller's earth into the water. Cider mill refuse, slag from the iron forges and shovel works and pollutants from cotton mill bleaching greens all found their way into the river as well. During the nineteenth century, among the most notorious producers of pollutants was one of the town's largest employers, the Star Mill. The mill discharged soapy, grease-laden wastewater at a probable rate of between three

Nemasket River at East Grove Street. Photograph by Rose Standish Pratt, circa 1912. The image hints at industry's harmful impact on the river. *Author's collection.*

hundred and five hundred pounds of grease for every 100,000 gallons of water directly into the river, along with wool waste and synthetic dyes, creating what must have been a foul grease-clogged river just above the Wading Place. Most detrimental of all, however, was the municipal sewage system constructed by Middleborough in the 1880s that had long-lasting effects on the health of the Nemasket and the future of the herring.

As early as 1883, the town had contemplated the construction of a public sewer system to serve Middleborough center, and by the mid-1880s, a system designed by noted hydraulic engineer Percy M. Blake of Hyde Park, Massachusetts, was operational, including an outfall located in the rear of the Wareham Street Municipal Light Plant, just a short distance downstream of the fishladder. There the town of Middleborough began discharging raw, untreated sewage directly into the river in what was to have been a temporary measure. The practice continued for more than sixty years.

In 1900, so offensive were Middleborough's sewage discharges that they were found to be compromising the water supply of the Massachusetts State Farm at Titicut in Bridgewater that drew its water supply for 1,200 people from the Taunton River just downstream from its confluence with the Nemasket. A committee established by Middleborough to investigate methods to treat its sewage saw "no pressing need of a change from the present mode" despite contamination of the state farm's supply. Similarly, Middleborough's representatives at the statehouse in Boston, Senator David Gurney Pratt and Representative William A. Andrews, in 1902 opposed a general pollution bill relative to inland waters because it would have negatively affected operation of Middleborough's sewer system by prohibiting direct discharges into the river.

By 1910, the polluted condition of the Nemasket had become intolerable for residents living close to the river. A combination of foul odors, unsanitary conditions and repeated malarial outbreaks prompted Muttock residents to petition for improvements to the river, primarily in the form of a reconstructed Muttock dam. These concerns were validated by the state board of health, which acknowledged that "the Nemasket river has been used as a place of disposal of the sewage of the town of Middleboro for many years. The river is also polluted by manufacturing wastes, and its condition below the sewer outlets during the year 1910 has been more objectionable than in any previous year."

Yet the circumstances and degraded condition of the Nemasket River evoked little concern locally. In 1913, when the state board of health again described the Nemasket as "badly polluted by the discharge of sewage and manufacturing wastes into the river," the *Middleboro Gazette* buried the story in the lower left-hand corner of the paper under the seemingly dismissive headline, "Next!" The Middleborough Board of Health simply refused to consider the situation at Muttock. Meanwhile, so bad had the condition of the Taunton River become that Taunton was ordered to desist from dumping sewage into that river by December 1, 1913.

The situation continued to worsen each year until 1919, when the commonwealth proposed investigating pollution in the Taunton River watershed, including the Nemasket River, spurred likely in part by Brockton's acknowledgement that it had polluted the Taunton River by discharging untreated sewage into it. Those Middleborough residents concerned that the proposed investigation was but the opening salvo of a larger campaign by the state department of health to gain control of the Nemasket and Taunton Rivers were adamantly opposed to the inquiry. Middleborough's representative in Boston, Morrill S. Ryder, in late May 1919 became "mightily keyed up in opposition to the proposed investigation" and sought to defeat it. Ryder's opposition was described as a "single-handed fight against a strong committee on public health, backed by the still stronger state department of health," a characterization bound to win him support in Middleborough, where resentment of perceived state interference traditionally ran high.

Nemasket River at Muttock, late 1800s. In 1910, when the Massachusetts Board of Health issued a damning assessment of the river's polluted condition, few took notice. *Author's collection.*

Although Ryder remained "fully loyal to his town," during the second day of debate, he was put on the defensive when he was unable to statistically refute charges that "the Nemasket river and the town of Middleboro were the worst offenders in the state, and that he was trying to help the Middleboro manufacturers [with whom he had close ties], regardless of the serious pollution which they caused." Those familiar with Ryder's connection with the varnish-making industry of Middleborough found the allegation believable, and support for Ryder quickly withered.

As the community had suspected, investigation of the condition of the Nemasket and Taunton Rivers resulted in draft legislation in 1920 by the Massachusetts Department of Health "for the protection of the public health in the valley of the Taunton river." The proposed law sought to eliminate pollution from the Taunton River watershed (the bulk of which was attributed to municipal sewage discharges by Middleborough, Taunton, Bridgewater and Brockton) by prohibiting "the entrance or discharge into any part of the Taunton river or its tributaries of sewage and of every other substance which, in the judgment of the department may be injurious to the public health or tend to create a public nuisance."

Acting in concert with the other Taunton River communities, Middleborough worked to defeat the bill, fearing that it would strip it of its control over the Nemasket River. When the bill was reported on unfavorably by the legislative committee to which it had been referred, it died; municipalities and manufacturers remained free to discharge pollution into the region's rivers. Ironically, it was the *Brockton Enterprise* and not the local newspaper that recognized the tragic consequences that the bill's defeat would have on the polluted Nemasket: "So the door is now open and offenders seem to be in the way of an overhauling for polluting a river which ought to be as pure and pellucid as its name is poetic."

ECONOMIC CONSEQUENCES OF DECLINE

Regardless of its causes, the decline in the Nemasket herring fishery was apparent in the dwindling sums it realized at the annual March auction. In January 1910, James A. Thomas gave voice to local dissatisfaction with the state of the run and what he perceived as the town's lack of vigilance in caring for it when he told the Middleboro Business Men's Club that the fishery "ought now to be three times as large. Middleboro has one of the best fish rivers in the state if it were properly cared for." Further affecting the fishery's income was the fact that funds from the communities downstream on the Taunton River that were used to help Middleborough finance the cost of maintaining the fishery were not always forthcoming during this period, due to declining Taunton River catches. This only aggravated the financial condition of the Nemasket fishery, particularly after these payments lapsed altogether in the depressed economic climate of the early 1930s. The failure to make more than token payments prompted Middleborough in 1939 to seek payments dating back six years and totaling $780 from seven Taunton River communities.

Bids for the Nemasket privilege throughout the early 1900s remained low, not only because of growing uncertainty about the viability of the run but also because of the volatility of the market for herring. Although herring were once considered valuable commodities, demand for the fish fluctuated greatly following 1900, as old markets evaporated and new markets proved sometimes difficult to secure. The *Middleboro Gazette* in 1904 noted the consequence of such changes in market conditions:

> *The herring fishery in town will not be a brilliant success this year. At the [Star] mills fishway from 4,000 to 5,000 alewives have been taken*

Empty fishing boat, Star Mill pool, 1950s. The image is evocative of the Nemasket fishery's twentieth-century decline. *Middleborough Public Library.*

daily, but there is no market for them. One lot of 8,000 fish recently sent to Boston netted but about $4. The more general use of herrings caught in salt water for bait is one explanation of the falling off in demand and price, and the large shipments formerly made to the West Indies have also fallen short.

In 1910, for the first time in many years, Nemasket River herring were pickled for shipment to the West Indies, with more than 350 barrels being shipped in mid-May. Yet this attempt to revive the former West Indian trade in Nemasket herring proved futile, and 1910 marked one of the last years when the fish were shipped to the islands as an inexpensive food source.

While less sought for human consumption with each passing year, herring came to be much favored as bait. In 1870, reporting on a large catch of white perch, the *Middleboro Gazette* revealed that the bait used had been juvenile herring, and it advised that "the disciples of Isaac Walton should 'go for' them" as so many on their return to the sea "are crushed and destroyed in different ways and do no one good." In 1887, Taunton River herring were "barreled as soon as caught and shipped to Boston and other places to mackerel fishermen" for use as bait. Herring were deemed "highly satisfactory" as line bait for ground fish, including cod and haddock, and "with the decline of the industry their exportations [to the West Indies] ceased, until, as bait for Boston and Gloucester fishing boats, their greatest demand came." Nonetheless, prices realized for local bait herring were much lower than those realized when the fish was pickled for export, and this tended to keep herring prices low. Additionally many fishermen found herring less desirable after mid-April, when more desirable bait became available. The one bright spot was the market for fresh herring, which still could fetch big prices. In April 1920, the first herring of the season were selling "at the extreme price of ten cents apiece," while a fish peddler at Wareham was getting even more, selling two for a quarter. Such prices, however, were transitory. Once the runs reached peak, prices plummeted to 50 cents per 100.

In 1925, "with the economy wave at its crest," the financial circumstances of the Nemasket herring fishery were considered so low and its prospects for the future so bleak that David R. Walker, at a special town meeting, moved that town counsel be instructed to present a bill to the Massachusetts legislature relinquishing Middleborough's rights in the local herring fishery. James A. Thomas opposed the measure, as did most others, and it fell flat. Not surprisingly, Middleborough could not be moved to part with its herring.

CHANGING CIRCUMSTANCES,
1921–1965

HERRING RUN AS TOURIST ATTRACTION

The decline in the herring runs during the late 1800s and early 1900s, and the consequent concern that was shown for the Nemasket fishery during the first decades of the twentieth century, were accompanied by a growing awareness of the herring's social and ecological role. The annual herring run became increasingly regarded as a particularly novel and picturesque occurrence, seen less as an economic activity and more as an attraction reflective of a distinctive New England tradition. In 1974, when noted Middleborough columnist Clint Clark of the *Middleboro Gazette* termed the annual herring run at Middleborough "a 'must see' happening" equivalent to the town's annual Fourth of July parade and fireworks, he linked the Nemasket run with two other local events celebrating uniquely American traditions. Clark's comment was clearly indicative of the consciousness, steadily growing since the start of the century, that the annual herring migration was an integral part of the community's identity and, as such, something worth witnessing.

The herring run at Muttock had early attracted residents, not only to capture the fish but also simply to witness the spectacle of the running herring. Jerusha B. Deane, a longtime resident of Muttock, recalled in 1914

Nemasket Street Bridge and Muttock millpond. Stereoview by W.H. Stodder, early 1880s. *Author's collection.*

that during the previous century the Muttock "weir was an attractive place, people coming from afar for the good fat herring which were caught in hand nets by thousands. Children coming from school liked to stop and 'see the herrings flupper,' as my small cousin used to say." Similarly, the *Middleboro Gazette* commented as early as 1873 that "it was just fun to see them catch them at the lower weir." Publication of a letter dated May 6, 1872, at Middleborough from Miss Mary L. Washburn, describing the scene of catching herring in picturesque terms by the *Boston Journal*, provided further evidence that the annual run was increasingly regarded as an attraction worthy of witnessing:

Here, also, the Taunton "alewives," the herrings, are just now. I wish I could picture the gay scene as the water dashes in a golden stream over the fall, and the fishermen throw their nets in the white foam and bring them up full of quivering silver fishes—then hold their nets reflectively in the air a moment, then again bring them up from the water and toss out their shining treasure.

In 1897, it was remarked that "at the opening of the season the days on which the fish are allowed to be taken are observed almost like holidays, and people come in carriages from far and near to see the 'herring run.'"

By 1905, when the local press remarked that "to see the herring 'run' is now one of the popular diversions and many visit the weirs," visits by spectators had become commonplace. The 1908 run reportedly brought many people to the Wareham Street fishway "to see the herring make their way over the falls," and the last week of April 1915 witnessed "scores" of people at the Wareham Street run. In an increasingly industrialized and urban world, this marvel of nature was attracting evermore numbers of observers, who watched from the sides of the ladders or the bridges above.

The growing popularity of the automobile and the prevalance of the street railway brought with them an increasing accessibility to nearby attractions, of which local herring runs at Middleborough, East Taunton, Pembroke, Bourne and elsewhere were one. In 1901, the *Boston Herald* described the herring's ascension of the East Taunton fish ladder as "one of the prettiest sights imaginable…. Here the fish can be seen plainly in bunches almost thick enough for one to walk across on their backs." Ten years later, East Taunton was noted as "a great Mecca for excursionists during the 'run.'"

The poor economic climate of the 1930s, followed by gasoline and tire rationing in the early 1940s, encouraged visits to herring runs that were both local and inexpensive, if not free. What could be more delightful than a spring motor trip accompanied by a picnic luncheon and a view of the running fish? Local advertising promoted the run to Depression-era visitors. "While in Middleboro Visit the Largest Herring Run in the World," urged one sign in 1936. To help educate visitors and make their visit more enjoyable, it was suggested in 1937 that signs be placed at each of the weirs in Middleborough in the hopes that "such signs might prove of value not only in preserving the fish from thoughtless destruction but also as an advertising medium to interested visitors."

Easter Sundays remained a popular time to visit the herring run. On that day, "people would show up dressed in their finery, seeing the fish so abundant, and under the false impression that they were easy to capture,

Nemasket Street Bridge and Muttock Dam, circa 1910. Residents gather at Easter to view the herring. *Author's collection.*

they would roll up their sleeves and pant legs of perhaps a newly purchased suit, wade into the water up to their knees, grabbing for the elusive fish until they were completely exasperated and totally drenched."

Following the war, visits to local runs remained high, keeping pace with the rise in private automobile ownership. In 1948, hundreds visited the Star Mill site to view the herring. "Folks arrived by auto, gazed at the fish for a while and then went their way." The following year, large crowds attended the arrival of the herring in early April. "Sunday was a big day for herring admirers. At each of the places where they can be seen best many came by car to look them over." In 1950, "there was a big audience and admission was free. It was the herring run. During the day yesterday, autoists from all over flocked to the fish runs at Muttock, the Star Mill and Wareham street, eager to get a look at them making their way upstream." The herring run even attracted out-of-state tourists, keen to witness the upstream migration.

By 1952, the number of visitors to the Nemasket herring runs reached "thousands," and traffic was so heavy that it created snarls requiring patrolmen to help direct sightseers. The last Sunday of March 1954 reportedly drew five thousand spectators to the Star Mill, and similar numbers would be quoted throughout the following decade.

The year 1952 also held the possibility of the herring run reaching a much wider audience with local rumor insisting that an unnamed television

East Taunton herring run, circa 1900. To profit from the herring's popularity, the run was later fenced and admission charged. Middleborough's runs were free. *Author's collection.*

Gathering the seine, Star Mill, 1950s. Herring harvesting presented a picturesque scene, attracting many spectators eager to see the fish. *Middleborough Public Library.*

network proposed to come to Middleborough in order to film the event. Although nothing appears to have come of this, a Providence television station did film the operation at the Star Mill site in April 1958. The run once more made it to television in 1966, attracting additional numbers of visitors to Muttock in April.

Not only television but also radio featured the local celebrity fish. In late April 1940, popular Boston radio personality "Caroline Cabot talked about the Middleboro herring run during her shopping tour broadcast on Station WEEI last Saturday morning, which probably attracted some of the Sunday crowd of sightseers."

While most of the tourists present at the runs came to view the herring as they struggled upstream, others found their fellow spectators themselves an equally fascinating object of interest, as the arrival of the fish led even the most rational individuals to adopt curiously atypical behaviors, as noted in 1951:

> *It hardly seems possible that adults could become so engrossed with this run of fish to act as they do. Some get their feet wet by reaching to catch them by hand. Others get their nice clothing mussed up and wet just for a chance to get a load of fish. While there is a frown if folks sought to help themselves generously to the fish, no objection is raised if visitors catch a few in their hands to carry away for their own eating.*

The Contentious 1940s Fishery

While growing in notoriety during the first decades of the twentieth century, the Nemasket herring fishery was simultaneously declining economically. Writing in 1908, James A. Burgess dismissed the suggestion that Middleborough could ultimately realize $1,000 annually on the lease of its fishery, as had been wildly overestimated. Burgess, however, was wrong. Following decades of decline, the Nemasket privilege witnessed the largest bids to date in its history during the 1940s, and in 1953, Middleborough would realize the single largest payment ever for its herring privilege: $11,000.

Such an unanticipated turn of events seemed unlikely even as late as 1943, when at the nadir of the war selectmen were stymied in their efforts to obtain a $2,000 minimum bid for the run, and one buyer threatened a $100,000 lawsuit when his own bid was not accepted. Bidders themselves "argued that the town should not be too anxious to get a high price since there was a food crisis coming, with plans already under way to can most of the fish, with the by-products to be turned into fish meal for cattle." James Creedon, covering the auction for the *Brockton Enterprise*, suggested the possibility of collusion among the potential buyers to keep the bidding low. The auction was characterized by a number of executive sessions, while "20-odd fish buyers, the press and a few casuals were herded out."

Some argued that given the war, the privilege that year should not be auctioned. During the previous war, in 1917, selectmen had declined to auction the fishery, instead selling the fresh fish to townspeople at the actual catching price (thereby helping to ease local food shortages) and corning the remainder for wintertime consumption in Middleborough and Lakeville.

Ultimately, the 1943 privilege was auctioned, and some fifteen thousand barrels were harvested, belying Belding's assertion that the Nemasket fishery had been exhausted. The season's catch was the largest ever hauled from the river, and "it clinched the claim that the river here is the prize river of the Atlantic coast for herring runs." Although bidders had assured Middleborough that captured herring would be earmarked exclusively for military commissaries and not marketed to the public, more cynical observers noted the cans of river herring that stocked local grocery store shelves, selling for fifteen to seventeen cents per can. James Creedon of the *Brockton Enterprise* estimated that some $100,000 had been realized by the 1943 lessees for what was considered a meager investment of $1,500.

This startling return on investment dramatically drove up bids the following year, when $8,600 was paid for the privilege by the Eastland Food Products Company of New Bedford, a figure reached after just four minutes of bidding and one that was "far and beyond anything which was ever offered." Although high, such prices for herring privileges were not entirely unknown. In 1920, G.O. Proctor of Gloucester paid Wareham the sum of $11,000 for that town's Agawam River privilege, five times the previous record. The reason for Proctor's bid was not clear, as the Agawam run remained unreliable, a fact that triggered speculation in town regarding Proctor's motives. "Wareham wonders what Mr. Proctor knows about herring, which it doesn't know."

Despite elation over 1944's record lease fee, the sum proved problematic, as it artificially inflated the value of the Nemasket herring fishery. Discouraged by the failure of the fishery at the start of May 1944, large fish packers and processors like Eastland avoided the 1945 auction, at which smaller operators could not meet the inflated expectations of

Seine harvesting at the Star Mill. Photograph by Fran Wiksten, 1960. *Author's collection.*

Middleborough selectmen, who refused to consider bids less than $5,000. Allegations of buyer collusion once more kept prices low, and charges that selectmen were inflating the value of the run in order to drive up bidding did little to foster trust on either side. Ultimately, the selectmen's minimum bid was not met, and the fishery went unleased that year.

The 1947 fishery presented problems of its own when the lessee, the Neptune Food Products Company of New Bedford, demanded refund of its $1,550 payment after a poor run resulted in the company taking only 125 barrels. Neptune's reasoning was simple: "The Neptune Company bought the alewives, and there were no alewives." The distinction overlooked by the company was that it had purchased the *right* to take herring from the Nemasket and not the herring themselves. Fishing the Nemasket remained a gamble. Kenneth Orr of the Eastland Food Products Company in 1948 acknowledged that "the Middleboro stream is known as the biggest gamble of all the fish runs."

The 1948 auction proved noteworthy when, for the first time, the fishery privilege was sold on a per-barrel basis rather than at a flat rate, as had been the practice. Although Neptune was again the highest bidder for the privilege, selectmen unsurprisingly refused the bid on the grounds that Neptune had breached its contract with the town the previous year by not paying the Middleborough wardens. "We have considered your bid. We are not happy over the way you used us last year. Unless you make amends we will not accept your bid," Board of Selectmen chairman Manuel J. Silvia told Neptune. Warden George A. Barney, for his part, refused to deal with Neptune and walked out of the auction.

"DISAPPEARANCE OF HERRING IN MIDDLEBORO CREATES MYSTERY"

Steady inflation of the Nemasket fishery's value, accompanied by growing friction between selectmen and buyers during the 1940s, occurred in the context of mounting unease over the health of the run, particularly following 1944, when Middleborough first confronted what disappearance of herring from the Nemasket River would look like—after an initial steady run, the herring vanished from the river on the night of May 4. "Since last week there hasn't been enough herring or alewives going up the river to fill a kettle of fish," despaired one local newspaper. Ironically, up until the fourth, the river at Muttock had been "bristling with fins." Warden Barney attested to never having witnessed anything like it in his sixteen years of experience. Fishermen engaged to harvest the fish instead spent their days lounging in the sun or sleeping in the nearby bunkhouse. Although the disappearance proved temporary (the fish would return as usual in 1945), developments in May 1944 were clearly unsettling for the community, particularly when it learned of circumstances downstream.

Stories soon reached Middleborough from East Taunton, where dead alewives littered the Taunton River's banks and floated in the river.

Middleborough's lessee that year, the Eastland Food Products Company, soon confirmed these reports, as did the *Middleboro Gazette*, which described the scene downriver: "On the main river dead fish, yellow bellies up, were floating on the surface like autumn leaves, and were scattered along the banks." Not surprisingly, offensive odors accompanied the die off nearly immediately. So numerous were the dead fish that soldiers from Camp Myles Standish in Taunton were dispatched to gather them and bury them along the banks of the Taunton River in order to prevent any potential health issues.

Those fish that did manage to survive appeared disoriented. Hundreds entered the Forge River and Dam Lot Brook in Raynham, streams that they had been unaccustomed to entering, previously having done so "only once when a broken dam upset natural conditions." A reporter for the *Middleboro Gazette* witnessed the pathetic scene at both locations:

> *Dead fish on the banks of both streams, some of them eight or 10 feet from the water, and Dam Lot brook was crowded with dead, dying and live fish, the live alewives seeming to be milling around as though they did not know just what to do or where to go, and shoving their dead companions aside.*

Various suppositions as to the cause of the situation circulated, with the *Middleboro Gazette* asking, "Poison or Heat the Cause?" Observers at Taunton concluded that either pollution of the Taunton River or the chemicals that had been added to it to combat two years of offensive odors arising from it were to blame for the die off. Perhaps not incoincidentally, Middleborough selectmen in November 1943 finally had raised the issue of the Taunton River's polluted condition with the state department of Marine Fisheries. Selectman William G.L. Jacob "was among the more insistent of the selectmen that the alleged pollution of the river in Taunton might have ill effects on the fish," and he claimed that both the alewives and the herring gulls that fed on them had been "poisoned."

At the time of the 1944 die off, an investigation was conducted that included analysis of the dead fish taken from the Taunton River, samples of river water from various points and a survey of the industrial operations discharging into the river. The final conclusion was that "exhaustion of oxygen in the river waters due to sewage and industrial pollution" had caused the massive die off in the Taunton River and had brought an abrupt halt to the run in Middleborough. Six companies, including the City of Taunton's own sewer department, were cited as the river's most egregious polluters. The Massachusetts Division of Sanitary Engineering called for an immediate reduction in the level of discharges detrimental to the fish and threatened legal action in the event that the companies failed to comply voluntarily.

Although the *Middleboro Gazette* chastised the state, whose actions in confronting the issue of pollution in the Taunton River it characterized as "dilatory," it failed to hold the local community responsible for its own misues of the Nemasket River. Some residents, however, were awakening to the need for a more critical examination of Middleborough's environmental record, and in the experience of 1944 the *Gazette* ultimately did see a cautionary tale, rhetorically asking readers, "And may there not be a hint for Middleboro in what has taken place? The Nemasket river is the disposal plant for our sewage."

Because of issues like the Taunton River, the Massachusetts Department of Health eventually acquired authority to order communities to cease pollution of rivers and streams and compel them to construct sewage treatment plants. Although Middleborough first seriously considered construction of a sewage treatment plant in 1938, World War II postponed any decisive action, and it was only following the war, with the tightening of the state's health laws and a threat of legal action against the town by the Massachusetts Board of Health, that Middleborough finally moved. When Middleborough's sewage treatment plant became operational in November 1950, the town's municipal wastewater discharges into the Nemasket River were treated for the first time ever, a positive development in part attributable to the herring.

THE MANNER OF FISHING

Early in the century, the work of harvesting herring was done by men with hand nets who scooped the fish from the river and loaded them into wagons or, more typically, wooden barrels of two hundred pounds' capacity. As late as the early 1930s, this was still the process followed, as remembered by Ralph Maddigan Jr., who in 1947 recalled "the old time fishing place in a stream [behind the Star Mill], which had a box where the fish gathered and…were hand netted out." Maddigan's cousin, James F. Maddigan Jr., who resided on the farm above the Star Mill privilege, also recalled the process:

As far back as I can remember, the fish were scooped up in large hand nets and deposited in barrels set in a large box, perhaps 8 feet by 10 feet in area;

any that missed the barrels in their loading could then be recovered and re-barreled. From here they were either hauled away, or moved to the herring house for processing.

Before the fish were packed into barrels, the men seining the river sorted the fish; anything other than herring and shad—such as perch, pickerel, eels, lamprey, bass, trout and turtle—was tossed back into the river.

Development of power-operated seines during the second quarter of the century drastically altered the harvesting process, as noted by the *Middleboro Gazette*: "When the seine is pulled the fish are brought out in a huge net, elevated by engine power, and dumped into trucks, which are then weighed, and in that manner the number of barrels per truck is established. They are hurried away in these huge trucks, principally to canneries to be processed as food." This new manner of fishing is described in the following passage that reflects the typical harvest as conducted during the 1940s and 1950s:

For years after the fish have entered the pool, a net has been set, with a rope rigging so that it can be gathered in and all the fish brought together at the base of the loading platform [at the edge of the river]. *A hoist, powered by a stationary engine, has lifted the fish into waiting trucks to get them on their way to processing.*

The fishermen would set their net across the river, below the dam, and leave it overnight. The next day they would wade into the stream and begin drawing in the net towards the small dock, on which was a winch, powered by an automobile engine. A small purse net, suspended from a boom, was dipped into the river net, to be hoisted full of fish. It was swung, either over barrels or a truck, and the catch released by pulling a rope at the bottom of the purse net.

During the 1954 harvest, a novel means of taking the fish was introduced by the Cundy Harbor Fisheries of Maine. While the fish were still netted and the ends of the net gathered together, a suction hose was inserted into the net, and the fish were removed through means of the hose and a water pump powered by a twenty-five-horsepower engine. The engine was "mounted in a truck, and the electrical gadgets and controls which operate it are in the truck. The truck has a temporary foundation of cement blocks." In order to power the machinery, Cundy paid an additional $250 to the Town of Middleborough for the Electric Light Department to run an electrical line to the loading platform:

Star Mill harvest, 1950s. A purse net draws herring from the seine in the rear of the former mill. *Middleborough Public Library.*

The fishing pool has a net spread, and workers in a boat service it. The lines of the net are then pulled taut and the net brought toward shore with the top floating on huge cork floats. Then the suction hose is lowered into it…

When the motor starts…it picks up the water and the fish from the river and they pass through an eight-inch rubber hose, which elevate[s] them about 15 feet, and the water and fish drop into a large box. There a worker can grasp the trout, perch and pickerel and an occasional bass, which make their way along the stream with the herring.

Efforts are made to hand pick all but the herring from the box. They are released down a sluice-way to a waiting truck. The other kinds of fish are restored to the water, apparently none the worse for their trip through the siphoned water which brings them up.

While the suction of alewives through a hose helped streamline the harvest and was deemed more efficient, it failed to work as expected. The suction machine mangled many of the fish, in the process destroying any value they may have had for any use other than fertilizer.

Despite the facility with which the herring could now be captured, the new methods had a number of drawbacks, including the depletion of all fish species in the river. In the past, fishermen had returned all fish other than herring to the river, but with new harvesting methods, time was no longer taken to sort the fish. All species went into the truck, depleting the stock of fish in the river. The means of transporting the fish, particularly in the early days of motor transport, also proved problematic at times. In 1939, the tailboard of a truck transporting the fish broke open, strewing the roadway with dead herring. "Passing autos skidded on them, and highway employees had to report at night to push them off the road and to sand it to make it safe for traffic." A similar mishap occurred the following year, when herring fell out of a hole in a truck and were deposited along North Main and Center Streets before the situation was noticed:

Autos proceeded to squash them. Some store keepers, figuring they might become fragrant, went out with boxes and picked the ones up in front of their store, while some more folks seeing a chance to get fresh herring without much effort picked them up and took them home.

Undoubtedly, those last ones were old-time Yankees.

BAIT, GURRY, CAT FOOD AND PEARLS

The purposes for which river herring were bought changed during the twentieth century as they became increasingly unappealing for human consumption except when food shortages threatened, as was the case during the two world wars. Fresh herring were either frozen or canned, and canned herring came to represent a substantial business, especially during World War II. Canning had the advantage, it was discovered, of softening the "needle-like bones" of the fish.

JONESPORT

BRAND

CONTENTS 15 OZ. AVOIR.

RIVER HERRING

SALT AND WATER ADDED

PACKED BY R. B. & C. G. STEVENS, JONESPORT, MAINE.

HERRING PATTIES

1 15-oz. can River Herrin
1½ cups mildly seasone
mashed potatoes, 1 teaspoo
grated onion, 2 teaspoo
chopped parsley, ½ teaspoo
salt, dash of pepper.

Remove center bone from fis
mash. Add other ingredient
mix well. Shape into mediun
size patties. Saute in hot f
one inch deep in frying pan
in deep fat 375° F., for abo
five minutes until brown o
both sides. Serve with chi
sauce. Serves 6-7.

In times when war did not drive the demand for edible herring, purchasers of the herring privileges often struggled to find a market for the fish. In 1940, although lessee James Ferreira of Plymouth believed that some 10,000 barrels could be taken from the river, he contended that there was no market for that many and proposed taking only 3,500 barrels, shipping the fish to Boston and New York.

Eventually, during the second quarter of the twentieth century, markets other than those for fresh and processed herring were to prove the principal means of disposing of the Nemasket catch that came to be used as bait for lobster and cod fishermen; gurry (fish offal fertilizer); fishmeal (a high-protein animal feed supplement produced by cooking, pressing, drying and grinding the fish); and cat food.

Herring had long been used by local recreational fishermen for bait, and commercial fishermen eventually came to realize the fish's value as well, beginning with mid-nineteenth century Gloucester fishermen. A century later, it was said that during 1952, lobstermen at Scituate and Cohasset, Massachusetts, had relied heavily on Nemasket herring as bait.

In 1940, New England fisheries processed some 24,000 cases of river herring, including some from Middleborough. By 1943, that number had reached 100,000. *Author's collection.*

In the mid-twentieth century, Nemasket River herring also came to be used in the manufacture of cat food and fertilizer. Reliable Fisheries of Plymouth, the successful bidder during a number of years in the 1950s and 1960s, transported its catch to cat food canneries in Maine for processing. Naturalist John Hay quipped in 1959 that "the most likely place to see indications of alewife now is on the stupendously bountiful shelves of a chain store, in the form of a can with a picture of a cat on it." The 1953 haul was transported by truck to Lubec, Maine, where it was processed into cat food, while the 1954 take was trucked to another cannery in Maine.

Contrarily, the catch taken by Cundy Harbor Fisheries in 1954 was processed for fertilizer. Hay seemingly found the trend away from human consumption of the fish and its mechanical reduction into fertilizer disconcerting: "'Reduction' is what they call it when the alewives are turned into fish meal, and in a sense perhaps they have been reduced, at least in our personal esteem. They now belong to a technical age with the rest of us."

The prices bid for the Nemasket fishery privilege fluctuated greatly following World War II, dependent as it was on both the gurry and cat

Glowing health,
Graceful beauty

Thanks to daily feeding of

PUSS 'n BOOTS

Wholesome nourishment balanced carefully in a daily diet that
contains every vital nutrient cats need – that's Puss 'n Boots.
Feed the Original Fish Formula if your cat prefers fish, the New
Meat Flavor if she likes meat. Both will keep your cat aglow
with radiant health, and a sunny disposition to match.

8 oz., 15 oz., and 26 oz. sizes

IMPORTANT

Cats have different food requirements from other
animals. If your cat likes fish, it needs a fish
food formulated just for cats: our Original Fish
Formula. If your cat likes meat, it needs a meat
food made just for cats: our New Meat Flavor.

Once a day—Every day—for lifelong nutrition

Coast Fisheries Division of The Quaker Oats Company, Chicago 54, Ill.

food markets. In March 1959, the Monument Fish Company of Provincetown secured the privilege with a record-high bid of $3.10 per barrel but withdrew the following month citing a $15.00-per-ton drop in the price for which it had expected to sell the fish to a Maine cat food cannery that had a glut on its hands. The privilege was subsequently leased to the Reliable Fish Company of Plymouth, which sold it to fertilizer producers in Gloucester and Woburn. "With six trailers on the haul, fishermen are apparently doing well at gurry price on the take," noted the *Middleboro Gazette* at the time.

In the 1960s, Middleborough herring were processed by Maine canneries and marketed under the Puss 'n Boots label, then America's most popular cat food brand. *Author's collection.*

In 1960, however, with both the gurry and cat food markets depressed, Middleborough considered not leasing its herring privilege. Discussions with the Reliable Fish Company, the lessee for the previous few years, however, encouraged it to do so. Yet there was a substantial difference between the bid offered in 1959 of sixty-five cents per barrel and that in 1960 of just fifteen cents, a drop attributable to the decline of the fertilizer market that was in its most depressed economic state in twenty years. Consequently, the catch during the early 1960s was hauled to a cat food manufacturers in Lubec, where it was processed into cat food and marketed under the Puss 'n Boots label.

Perhaps the most unusual use to which herring were put was the manufacture of artificial pearls. Guanine, the iridescent substance of the fish's scales that give the herring its sheen, was found to be extractable, and it was processed into a lacquer known as pearl essence, or essence d'Orient, that could be applied to glass beads to create imitation pearls, sometimes known as "Priscilla pearls." Industrialization of this process in the 1920s saw the use of pearl essence in the production of not only pearls but also toilet articles, buttons, automotive paint and cosmetics. Its use was short-lived, being rapidly replaced with synthetic substitutes following World War II.

"Holiday for Hooligans"

One continuing challenge in protecting the Nemasket herring during the mid-twentieth century were poachers and vandals, who continued to plague the fishery as they had for nearly three centuries. Fortunately for depredators of the run, even the most staid, law-abiding residents tended to look the other way when herring were involved. In 1937, the *Middleboro Gazette* noted, tongue-in-cheek, "Of course it's against the law to remove the alewives but if you drive an automobile such a minor offense as breaking laws should not trouble your conscience."

Perhaps more troublesome than the unauthorized taking of a few fish was the harassing of the up-running fish by means of throwing rocks into the fishladders. In March 1946, Middleborough fish warden George A. Barney encountered a group of boys engaged in the practice whose names he took and to whom he offered a stern lecture. "What really gripes Barney, however, is to catch an adult throwing rocks into the fishway," noted the *Middleboro Gazette*. Rock-throwing remained a constant pastime, and in 1950, juveniles were noted taking "delight in tossing rocks at the fish or otherwise annoying them." In the absence of a fish warden that year, local police assumed the task of patrolling the run. The 1952 run saw young boys with bows and arrows attempting to shoot the fish, an effort that was stopped by former Middleborough police chief Alden C. Sisson, who had been named herring officer by the selectmen.

As the herring runs attracted larger crowds each spring, police were on hand to prevent poaching and molesting of the fish. "In the past folks have decided to help themselves generously to the fish," noted one correspondent in 1949, "but town officials frown on the practice." In April 1950, a group from Norwood, Massachusetts, was noted as having its net confiscated by the police, "to add to several more which have been taken away from prospective fish-catchers." Nets and fishing gear were also reported as having been confiscated from "juveniles" in 1953.

Although young boys had always historically tried to catch the fish or more seriously molest them on their upstream run, commentary during the 1950s cast this mischief in terms of juvenile delinquency, then an emerging social issue:

> Small boys…were on hand at each of these places [Muttock, the Star Mill and Wareham Street] trying to catch herring. Some reached with their hands. Others sought to knock them out with sticks…. And Police Chief Charles Rogers is alerted to keep young America under control and protect the fish.

The well-loved Clint Clark, long-running columnist of the *Middleboro Gazette*, was equally scathing in his sentiments after witnessing boys at Muttock capturing the fish and throwing them back into the millpond, as well as other stunts:

> It is the unruly abuse of the fish which arouses many Middleboroeans. I hope that between now and next spring, the town fathers will see the wisdom of making the herring run an orderly tourist attraction, rather than a holiday for hooligans…. As it is now, there are no signs, and no sources of information. But there is disorder, ignorance, and kids competing to see who can abuse the most fish. Surely we, the Conservation Commission, Selectmen, and the Oliver Mill Park Committee, can do better than that!

Clark's own experience patrolling the river at Muttock prompted him to advise others against it "unless gifted with a great deal of patience and fortitude," citing "belligerent parents who didn't see any harm in their kids wading into the stream and hurling fish onto the bank." Little respect was shown the wardens, and three boys taunted Clark, "What are you—some kind of cop? Where's your badge?"

Star Mill and Assawompsett

Throughout the 1940s and 1950s, the Nemasket River, in the rear of the former Star Mill, served as the sole municipal fishing site for the town of Middleborough, with the town purchasing the Star Mill dam in July 1944 "for [the] purpose of protecting the fishing pool and conservation of herring in the river." The town, however, appears to have neglected the wooden dam, so much so that by 1954 it had largely fallen away, substantially lowering the level of the millpond behind it.

Like the dam, the natural fishway at the Star Mill also had a checkered history. It appears that 1943 was the final year in which the fishway was utilized, and in 1944, it was described as being devoid of water. In response, the Massachusetts Department of Conservation's Division of Marine Fisheries agreed to build a temporary fishway that in 1944 was "constructed through the dam to modify the flow of water, which, in spite of low water in the river, boils through the fishway with considerable violence." In 1948, Fish Warden George A. Barney indicated that the fishway needed to be dug out and reinforcing stones placed in it to shore it up. Barney complained of boys who "throw stones into the fishway and block it up." At the time, Barney was in favor of the proposal by the Division of Marine Fisheries to fund up to 80 percent of the cost of a new fishway on the expectation that Middleborough would fund the remainder, yet this the town would not do. Consequently, in March 1949, the fishway was reported as still requiring clearing and filling in order to provide passage for the fish.

Given these issues, a 1955 town-appointed committee of three—consisting of Rhodolphus P. Alger, Robert F. Howes and Albert T. Maddigan and charged with reviewing the Star Mill site and "the advisability of keeping the alewife fishery at this or any other location"—recommended relocation of fishing operations to Wareham Street.

Part of the rationale for relocating the the municipal seining pool was that at the Star Mill it was located on private property, owned since 1928 by the

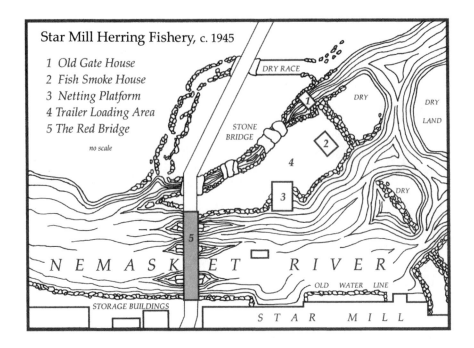

Star Mill Herring Fishery, c. 1945

1 Old Gate House
2 Fish Smoke House
3 Netting Platform
4 Trailer Loading Area
5 The Red Bridge

no scale

DRY RACE

DRY

DRY

LAND

STONE
BRIDGE

DRY

N E M A S K E T R I V E R

OLD WATER LINE

STORAGE BUILDINGS

S T A R M I L L

Maddigan family, who permitted the taking of herring from its property, although lessees of the town's privilege were required to pay a tariff on each barrel of herring taken for use of the family's "red bridge" (so named for the color of its painted rails), land and a small structure with a telephone known as the "office building." The fee was generally three cents for each barrel taken from the river and was meant to compensate for the large number of heavy trucks passing over the bridge.

One of the chores of Herb Willis, who resided during his youth in the early 1930s on the farm above the pool with the family of his uncle, Frank Maddigan, was to count the number of barrels of fish taken. Willis recalled that the fish

> would come up the river in the spring to spawn, and would gather at a dam on the farmland and pass upstream on a ladder stream network. Many of the herring were harvested and made into fish oil and fertilizer. One of my jobs while on the farm was to count the number of barrels of herring caught each day. My count determined the official count for the day's catch.

Although abandonment of the Star Mill location was recommended in 1955, some residents were hesitant, so deeply associated was the site with

Star Mill privilege, 1950s. Reluctance to make improvements at the site was reflective of growing uncertainty regarding its future. *Middleborough Public Library.*

Lake Assawompsett, circa 1910. Excessive drawdown of the lake in 1949 and 1957 trapped juvenile herring with no means of entering the Nemasket. *Author's collection.*

herring. Before World War II, considerable publicity was given to the herring run at the Star Mill as "the largest one in the world" (or so the sign once located there proclaimed).

Another problem adversely affecting the Nemasket herring at this time was the restriction of the natural outflow from Lake Assawompsett; this resulted in a diminished flow of water running in the river in summer, when herring fry made their return to the sea. The situation of low water in the Nemasket worsened during the 1940s, caused by New Bedford's excessive drawdown of Little Quitticus Pond and the sandbagging of Lake Assawompsett at the head of the river. So low, in fact, was the water during the late summer of 1949 that Lake Assawompsett's waters flowed in reverse into Pocksha Pond rather than out of it, with regional headlines trumpeting the "Odd Movement of Low Water." "In some places there was no water running in the upper reaches of the river," reported the *Brockton Enterprise*. Although this unique phenomenon was explicable, it was one that greatly troubled Middleborough residents. The newspaper noted that "it is not related if this has happened before," but it was unlikely.

In November 1957, the Nemasket River once more reversed its flow when midsummer drought conditions prompted New Bedford and Taunton water officials to close the Assawompsett dam, leading to an excessive lowering of Lake Assawompsett over the subsequent months. So disturbing was the circumstance of the river's reversed flow that the New Bedford *Standard Times* felt compelled in its story to employ capital letters, announcing that the Nemasket "HAS REVERSED ITS FLOW." Standing at the gatehouse, Middleborough officials "observed the river to be backed up against the planks at a level one foot higher than in the lake." Concerns were expressed for juvenile herring still remaining in the lake, and fears were expressed that the onset of winter and an early freeze would kill them. As a result, planks were withdrawn from the Assawompsett dam and the waters equalized, permitting the fish to pass.

AN ABANDONED PRIVILEGE

Despite the challenges faced by the Nemasket fishery at mid-century, the town's herring privilege continued to be auctioned, and the Star Mill site was used as the municipal seining pool until 1965, when the river channel there

Star Mill harvest. Photograph by George Morse, 1910. Such scenes had long since passed by 1965, when commercial herring harvesting in Middleborough ended. *Author's collection.*

was relocated to accommodate physical expansion of the Winthrop-Atkins Company, which since 1944 had occupied the former Star Mill complex. One consequence of the project was the burial of the municipal fishing pool under several layers of fill, necessitating identification of a new harvest site.

Although Wareham Street offered an alternate location for an industrial fishing operation, problems there persisted. In April 1940, shortly before commencement of the run, improvements on the Wareham Street ladder were abruptly abandoned. Although flashboards had been removed and work had been started on enlarging the channels in the ladder's cross-walls, the National Youth Administration boys engaged to perform the work were pulled off the project, leaving the cement "boxes" that made resting pools ineffective, requiring the installation of temporary flash boards and gates. Eventually, the ladder was completed, but occasional high water levels in the Nemasket River made its design deficiencies evident when water spilled over the shallow walls and poured in force through the run, making it nearly impossible for the herring to ascend. Although in time the sidewalls of the upper pools were raised to accommodate high water, the situation merely pointed out continuing difficulties in constructing a suitable ladder for the Wareham Street site.

Wareham Street ultimately was ruled out as a commercial harvesting site in the summer of 1965 following a review by both Middleborough and

Lakeville selectmen. Meanwhile, revitalization of the river at Muttock as Oliver Mill Park made relocation to that site equally unlikely, as conflicting industrial, ecological, historical and recreational interests promoted their own particular visions of the site. By March 1966, plans for relocation of the municipal fishing pool were described as being at a standstill, and although the selectmen set the date of the annual herring auction for March 28, the privilege was not sold that year for lack of a site at which to harvest the fish. Although not expected at that time, 1965 would be the final year in which Middleborough would auction its herring privilege, and a centuries-old tradition would disappear without notice.

AN EMERGENT STEWARDSHIP, 1966–2014

Awakening an Environmental Consciousness

Although no longer valued for its commercial importance after 1965, the Nemasket herring assumed a new environmental significance concurrent with a reassessment of the ecological heritage of the Nemasket River, which the Middleborough Conservation Commission in 1981 termed "Middleborough's most valuable and scenic resource." Work was undertaken to reconstruct the fishladders along the river's course. In 1968, a Denil fishway, the first of its kind in Massachusetts, replaced the aging fishway at Assawompsett, and a new ladder at Oliver Mill Park in 1983 was constructed following a problem with fish being stranded because of the old ladder.

Municipal and regional planning initiatives—including completion of Middleborough's first open space plan in 1974, efforts to establish a Nemasket River Environmental Corridor, establishment of a Nemasket River subcommittee by the Middleborough Conservation Commission and formation of a Natural Resources Preservation Committee in 1988—directly or indirectly had the goal of protecting the Nemasket herring. Regional cooperative initiatives including Middleborough, Lakeville, Freetown, Rochester, Taunton and New Bedford also focused on the Assawompsett Pond Complex watershed as an area of critical concern. New Bedford worked to

Wareham Street ladder, 2014. *Photograph by Michael J. Maddigan.*

improve the situation of herring in the upper ponds by limiting access to Little Quitticus, where the fish might be drawn into the city's water supply intakes, and improving egress from Great Quitticus for juvenile herring in the fall.

Most significant of all was the establishment in 1996 of the Middleborough-Lakeville Herring Fisheries Commission, only one of two of its kind in the commonwealth having direct control over its local fishery. The commission was tasked with monitoring the river, maintaining its fishways and regulating the catch. These efforts helped contribute to an increase in the Nemasket herring population, which peaked in 2000 when an estimated 2 million fish passed up the river to spawn.

The commission also helped foster exhausted herring runs throughout New England. As early as 1872, alewives had been taken from the Nemasket River in order to help build runs elsewhere. In that year, the town of Halifax paid twenty-one dollars for nine hundred herring that it took to stock its ponds in the hopes of establishing a run there. The attempt met with disastrous results. Of the nine hundred fish purchased, only seventy-five survived the journey from Middleborough to Halifax. Nevertheless, a century and a quarter later, Nemasket River herring were still being taken to help restore runs throughout New England, including the Cocheco River in New Hampshire; the Concord, Ipswich and Town Rivers in Massachusetts; and the Pawcatuk, Saugatucket, Annaquatucket and Ten Mile Rivers and Buckeye Brook in Rhode Island.

Decline

The spring of 2005 witnessed a startling 80 percent decline in herring populations statewide, including at the Nemasket River. In response, Massachusetts Division of Marine Fisheries in November imposed a total ban on the harvest, sale and possession of river herring through the end of 2008. While not obligated to comply with the ban, the Middleborough-Lakeville Herring Fisheries Commission voted to support the restrictions in an effort to rebuild the run, as well as to avoid potential conflict with the law that criminalized the possession of river herring.

As in the past, various theories were posited for the decline, including unusually poor weather, natural predators such as striped bass and seals and commercial fish trawling that captured enormous numbers of river herring as a bycatch. The decline in regional alewife populations historically was without precedent and was particularly disturbing given the herring's ecological role as a link in the riverine food chain.

As of this writing, the moratorium remains in place, as does a 2012 federal ban on harvesting river herring along the entire eastern seaboard. Both will remain in force until sustainability of river herring populations can be secured. While increases have been noted in the herring populations locally, the future of the Nemasket River herring remains precarious and can no longer be taken for granted.

CONCLUSION

No simpler pronouncement of the importance of the Nemasket herring's impact on Middleborough's development or of the interconnectedness of nature and culture may be made than that by the Herring Alliance, which in 2007 stated, "The river herring fishery connects us to our Nation's past." Herring have been a contributor to the social organization and historical development of communities throughout the region, with Middleborough in particular providing the example of how a small New England community has been utterly transformed by the fish.

Middleborough was fortunate that well into the twentieth century, river herring were valued economically, and this accordingly afforded a level of protection for the fish. By the time there was no longer a perceived economic worth, due to the declining number and uses for the fish, herring had acquired an ecological value as an important link in the food chain and as an indicator of a healthy ecosystem. In the meantime, herring had become closely wedded to a uniquely New England and specifically southeastern Massachusetts identity that underscored the fish's historic role. Whether as a cause for the expansion and eventual sedentism of Native American settlement, a factor in the survival and growth of the earliest English colonies, a determinant of the town's present economic geography, a contributor to the development of our social organizations and communal culture, a deterrent to unfettered industrial expansion, a resource for public welfare and municipal tax relief, a reason for reversing the degradation of the Taunton River watershed or a rationale for our present environmental awareness, the river herring, simply while we were not looking, shaped who we are as New Englanders.

In the process, Middleborough developed a sense of obligation to protect the fish as a vital historical and ecological resource. This stewardship, underpinned by an understanding of the cultural ecology of the Nemasket herring fishery, remains critical to the long-term survival of the Nemasket herring and has wider implications for the successful survival and hopeful reestablishment of herring runs elsewhere. To paraphrase Thoreau, by knowing the history of the Nemasket River herring and understanding the role the fish has played in our social and cultural development, we and others cannot but be affected favorably.

BIBLIOGRAPHY

ARCHIVAL RESOURCES

Massachusetts Historical Commission.
Massachusetts State Archives.
Massachusetts State Library.
Middleborough Historical Association.
Middleborough Public Library.
Plymouth County Registry of Deeds.
Taunton Public Library.

MIDDLEBOROUGH TOWN REPORTS

Middleborough Town Reports, 1866–2013, Middleborough Public Library.

NEWSPAPERS AND PERIODICALS

Boston Globe, Boston, Massachusetts.
Brockton Enterprise, Brockton, Massachusetts.
Brockton Times, Brockton, Massachusetts.
Middleboro Gazette, Middleborough, Massachusetts.
Middleboro News, Middleborough, Massachusetts.
Middleborough Gazette and Old Colony Advertiser, Middleborough, Massachusetts.
Namasket Gazette, Middleborough, Massachusetts.
New York Times, New York City, New York.
Old Colony Memorial, Plymouth, Massachusetts.
St. Paul Daily Globe, St. Paul, Minnesota.
Taunton Daily Gazette, Taunton, Massachusetts.

PUBLISHED SOURCES

Abbott, Katharine M. *Old Paths and Legends of New England, Saunterings Over Historic Roads with Glimpses of Picturesque Fields and Old Homesteads in Massachusetts, Rhode Island, and New Hampshire.* New York: G.P. Putnam's Sons, 1904.

Adams, James Truslow. *The History of New England in Three Volumes.* Vol. 1, *The Founding of New England.* Boston: Little, Brown, & Company, 1927.

Atlantic States Marine Fisheries Commission. *Stock Assessment Report No. 12-02 of the Atlantic States Marine Fisheries Commission: River Herring Benchmark Stock Assessment*, May 2012.

Atwood, Charles R. *Reminiscences of Taunton, in Ye Auld Lang Syne.* Taunton, MA: Republican Steam Printing Rooms, 1880.

Austin, Jane Gifford. *David Alden's Daughter and Other Colonial Stories of Colonial Times.* Boston: Houghton, Mifflin and Company, 1892.

———. *Standish of Standish: A Story of the Pilgrims.* Boston: Houghton, Mifflin and Company, 1889.

Baxter, James Phinney Baxter, ed. *Documentary History of the State of Maine.* Vol. 15. Portland, ME: Lefavor-Tower Company, 1910.

Baylies, Francis. *An Historical Memoir of the Colony of New Plymouth.* Boston: Hilliard, Gray, Little, and Wilkins, 1830.

Belding, David. *A Report Upon the Alewife Fisheries of Massachusetts.* Boston: Massachusetts Division of Fisheries and Game Department, Department of Conservation, 1920.

Bigelow, H.B., and W.C. Schroeder. "Fishes of the Gulf of Maine." *U.S. Fish and Wildlife Service Bulletin* 53 (1953): 1–577.

Bigelow, John. *Jamaica in 1850.* New York: G.P. Putnam, 1851.

Blaisdell, Albert Frank, and Francis K. Ball. *American History Story-Book.* Boston: Little, Brown, & Company, 1910.

The Book of Taunton: Illustrated. Taunton, MA: C.A. Hack & Son, 1907.

Boyd, Stephen G. *Indian Local Names, with Their Interpretation.* York, PA: self-published, 1885.

Bradford, William. *Bradford's History "Of Plimouth Plantation"* [1630–50]. Boston: Wright & Potter, 1898.

Bradford, William, and Edward Winslow. *A Relation or Journal of the Beginnings and Proceedings of the English Plantation Settled at Plimouth in New England, by Certain English Adventurers Both Merchants and Others.* London: John Bellamie, 1622. Reprinted. This work is commonly known as *Mourt's Relation.*

The Bridgewater Book: Illustrated. Boston: George H. Ellis, Printer, 1899.

Buesseler, Wendi. *The History of the Coonamessett River.* Coonamessett River Restoration Working Group library, Town of Falmouth, Massachusetts.

Carrier, Lyman. *The Beginnings of Agriculture in America.* New York: McGraw-Hill, 1923.

Carter, Susanna. *The Frugal Housewife.* New York: G. & R. Waite, 1803.

Ceci, Lynn. "Fish Fertilizer: A Native North American Practice?" *Science* 188, no. 4 (April 4, 1975): 26–30.

Chartier, Craig. "Welcome Back the Herring Program." *'Round Robbins.* Newsletter of the Friends of the Robbins Museum. Massachusetts Archaeological Society, Fall 2006.

Child, Lydia Maria Francis. *The Frugal Housewife.* Boston: Carter and Hendee, 1830.

Clapin, Sylvia. *A New Dictionary of Americanisms; Being a Glossary of Words Supposed to be Peculiar to the United States and the Dominion of Canada.* New York: Louis Weiss & Company, 1902.

Coffin, Charles Carleton. *Old Times in the Colonies.* New York: Harper & Brothers Publishers, 1880. Reprinted in 1908.

Coggeshall, Robert C.P. *The Development of the New Bedford Water Supplies.* New Bedford: Old Dartmouth Historical Society, 1915.

Cole, Samuel V. "Taunton—An Old Colony Town." *New England Magazine* 20, no. 1 (March 1896): 65–85.

The Compact with the Charter and Laws of the Colony of New Plymouth: Together with the Charter of the Council at Plymouth, and an Appendix, Containing the Articles of Confederation of the United Colonies of New England, and Other Valuable Documents. Boston: Dutton and Wentworth, 1836.

The Cook Not Mad, or Rational Cookery. Watertown, NY: Knowlton & Rice, 1831.

Course of Study for United States Indian Schools. Washington, D.C.: Department of the Interior, Office of Indian Affairs, 1922.

Cronon, William. *Changes in the Land: Indians, Colonists, and the Ecology of New England.* New York: Hill & Wang, 1983.

Davol, Ralph. *Two Men of Taunton: In the Course of Human Events, 1731–1829.* Taunton, MA: Davol Publishing Company, 1912.

Day, Anne Marjorie. *The Guiding Light: Pilgrim Tercentenary Pageant Play in Four Episodes.* Boston: Richard G. Badger, 1921.

DeBow, J.D.B. "The American Fisheries." *DeBow's Review: Agricultural, Commercial, Industrial Progress and Resources* 2, no. 5 (November 1866): 470–81. New Series, New Orleans.

DeVoe, Thomas Farrington. *The Market Assistant, Containing a Brief Description of Every Article of Human Food Sold in the Public Markets of the Cities of New York, Boston, Philadelphia, and Brooklyn.* New York: Hurd and Houghton, 1867.

Dix, Beulah Marie. *Soldier Rigdale: "How he Sailed in the 'Mayflower' and How He Served Miles Standish."* New York: Macmillan Company, 1899.

Doherty, Katherine M., ed. *History Highlights: Bridgewater, Massachusetts a Commemorative Journal.* Bridgewater, MA: Bridgewater Bicentennial Commission, 1976.

Drake, Samuel Gardner. *The Book of the Indians; or, Biography and History of the Indians of North America, from Its First Discovery to the Year 1841.* Boston: Samuel G. Drake, 1845.

Dwelley, Jedediah, and John F. Simmons. *History of the Town of Hanover, Massachusetts, with Family Genealogies.* Hanover, MA: Town of Hanover, 1910.

Eggleston, Edward. *A First Book in American History: With Special Reference to the Lives and Deeds of Great Americans.* New York: American Book Company, 1889.

Emerson, Lucy. *The New-England Cookery.* Montpelier, VT: printed for Josiah Parks, 1808.

Emery, Stephen Hopkins. *History of Taunton, Massachusetts, from Its Settlement to the Present Time.* Syracuse, NY: D. Mason & Company, Publishers, 1893.

Ewins, P.J., D.V. Weseloh, J.H. Groom, R.Z. Dobos and P. Mineau. "The Diet of Herring Gulls (*Larus argentatus*) during Winter and Early Spring on the Lower Great Lakes." *Hydrobiologia* 279–80, no. 1 (April 1994): 39–55.

Farmer, Fannie Merritt. *The Boston Cooking-School Cookbook*. Boston: Little, Brown, & Company, 1896.

———. *Food and Cookery for the Sick and Convalescent*. Boston: Little, Brown, & Company, 1904.

Flannery, Regina. *An Analysis of Coastal Algonquian Culture*. The Catholic University of America Anthropological Series No. 7. Washington, D.C.: Catholic University of America Press, 1939.

Forbush, Edward Howe. *Birds of Massachusetts and Other New England States*. Part I, *Water Birds, Marsh Birds and Shore Birds*. Boston: Massachusetts Department of Agriculture, 1925.

Freeman, Mary Eleanor Wilkins. *A New England Nun*. New York: Harper & Brothers Publishers, 1919.

French, Benjamin F. *Biographia Americana; or, A Historical and Critical Account of the Lives, Actions, and Writings, of the Most Distinguished Persons in North America; from the First Settlement to the Present Time*. New York: D. Mallory, 1825.

Goode, G.B. "The Use of Agricultural Fertilizers by the American Indians and the Early English Colonists." *American Naturalist* 14, no. 7 (July 1880).

Graham, J.J. "Observations on the Alewife, *Pomolobus pseudoharengus* (Wilson), in Fresh Water." Publ. Ontario Fish Res. Lab., No. 74., Ser. 62. Toronto: University of Toronto, 1956.

Greene, Sarah P. McLean. *Cape Cod Folks*. Boston: A. Williams & Company, 1881. Reprint, Boston: DeWolfe, Fiske & Company, Publishers, 1904.

Griffis, William Elliot. *The Romance of American Colonization: How the Foundation Stones of Our History Were Laid*. Boston: W.A. Wilde & Company, 1898.

Griffith, Henry S. *History of the Town of Carver, Massachusetts: Historical Review, 1637 to 1910*. New Bedford, MA: E. Anthony & Sons Inc., printers, 1913.

Hanna, William F. *A History of Taunton, Massachusetts*. Taunton, MA: Old Colony Historical Society, 2007.

Harland, Marion [Mary Virginia Terhune]. *Breakfast, Luncheon and Tea*. New York: Scribner, Armstrong & Company, 1875.

———. *Common Sense in the Household: A Manual of Practical Housewifery*. New York: Scribner, Armstrong & Company, 1873.

Harper's Encyclopaedia of United States History from 458 A.D. to 1905. New York Tribune edition. New York: Harper & Brothers Publishers, 1905.

Hay, John. *The Run*. Garden City, NY: Doubleday & Company, 1959.

Higginson, Reverend Francis. "New-Englands Plantation, or, A Short and True Description of the Commodities and Discommodities of that Country, Written in the Year 1629." 3rd ed. In *Mass. Hist. Soc. Coll.*, vol. 1. London: 1630.

Holton, Edith Austin. *Yankees Were Like This.* New York: Harper & Brothers Publishers, 1944.

Huden, John C. *Indian Place Names of New England.* Contributions from the Museum of the American Indian Heye Foundation. New York: Museum of the American Indian, 1962.

Hurd, D. Hamilton. *History of Plymouth County, Massachusetts, with Biographical Sketches of Many of Its Pioneers and Prominent Men.* Philadelphia: J.W. Lewis & Company, 1884.

Jenkins, Charles W. *Three Lectures on the Early History of the Town of Falmouth Covering the Time from Its Settlement to 1812.* Falmouth, MA: L.F. Clarke, Steam Printer, 1889.

Jewett, Sarah Orne. "River Drift-Wood." *Atlantic Monthly* 48 (October 1881): 500–10.

Josselyn, John. *New-England's Rarities Discovered.* London: G. Widdowes, 1672. Reproduced, Boston: Massachusetts Historical Society, 1972.

Kingman, Bradford. *History of North Bridgewater, Plymouth County, Massachusetts, from Its Earliest Settlement to the Present Time, with Family Registers.* Boston: Bradford Kingman, 1866.

Kissil, George William. "Contributions to the Life History of the Alewife (*Alosa pseudoharengus*) [Wilson], in Connecticut." PhD diss., University of Connecticut, 1969.

Krim, Arthur J. "Acculturation of the New England Landscape: Native and English Toponymy of Eastern Massachusetts." *New England Prospect: Maps, Place Names and the Historical Landscape.* Edited by Peter Benes. The Dublin Seminar for New England Folklife. Boston: Boston University Press, 1980, 69–88.

Lane, Helen H. *History of the Town of Dighton, Massachusetts, the South Purchase, May 10, 1712.* Dighton, MA: Town of Dighton, 1962.

Lanier, Sidney. *Poems.* Philadelphia: J.B. Lippincott & Company, 1877.

Lee, Mrs. N.K.M. *The Cook's Own Book.* Boston: Munroe and Francis, 1832.

Leonard, Mary Hall. "Old Rochester and Her Daughter Towns." *New England Magazine* 20, no. 5 (July 1899): 613–36. New Series.

Leonard, Mary Hall, et al. *Mattapoisett and Old Rochester, Massachusetts, Being a History of These Towns and Also in Part of Marion and a Portion of Wareham.* New York: Grafton Press, 1907.

Leslie, Eliza. *The Lady's Receipt-Book: A Useful Companion for Large or Small Families.* Philadelphia: Carey and Hart, 1847.

Lincoln, Joseph C. *Cape Cod Yesterdays.* Boston: Little, Brown, & Company, 1935.

———. *Mary-'Gusta.* New York: D. Appleton and Company, 1916.

Litchfield, Henry Wheatland. *Ancient Landmarks of Pembroke.* Pembroke, MA: George Edward Lewis, 1909.

Little, Eliza. *Directions for Cookery, in Its Various Branches.* Philadelphia: E.L. Carey & Hart, 1840.

Lutins, Allen, and Anthony P. DeCondo. "The Fair Lawn/Paterson Fish Weir." *Bulletin of the Archaeological Society of New Jersey* 55 (1999).

Maddigan, Michael J. *Star Mill: History & Architecture.* Middleborough, MA: Recollecting Nemasket, 2012.

Marble, Annie Russell. *Standish on Standish.* Boston: Houghton Mifflin Company, 1919.

———. *The Women Who Came in the Mayflower.* Boston: Pilgrim Press, 1920.

Milner, James W. "The Shad and the Alewife." *Harper's New Monthly Magazine* 60, no. 360 (May 1888): 845–57.

Mood, Fulmer. "John Winthrop, Jr., on Indian Corn." *New England Qaurterly* 10, no. 1 (March 1937).

Morris, Charles. *A History of the United States of America: Its People and Its Institutions.* Philadelphia: J.B. Lippincott Company, 1897.

Morton, Thomas. *New English Canaan or New Canaan.* Amsterdam, Netherlands: Jacob Frederick Stam, 1637. Reprinted, Boston: Prince Society, 1883.

Mullen, D.M., C.W. Fay and J.R. Moring. "Alewife/Blueback Herring. Species Profiles: Life Histories and Environmental Requirements of Coastal Fishes and Invertebrates (North Atlantic series) USDI Fish and Wildlife Service." *Biological Report* 82, no. 11.58 (1986).

Nanepashemet. "It Smells Fishy to Me: An Argument Supporting the Use of Fish Fertilizer by the Native People of Southern New England." *Algonkians in New England: Past and Present.* The Dublin Seminar for New England Folklife. Boston: Boston University, 1993, 42–50.

Nash, Gilbert. *Historical Sketch of the Town of Weymouth, Massachusetts, from 1622 to 1884.* Weymouth, MA: Town of Weymouth (Weymouth Historical Society), 1885.

Nelson, Gary A. *A Guide to Statistical Sampling for the Estimation of River Herring Run Size Using Visual Counts.* Technical Report TR-25. Gloucester, MA: Massachusetts Division of Marine Fisheries, Annisquam River Marine Fisheries Station, February 2006.

Odell, Theodore Tellefsen. "The Life History and Ecological Relationships of the Alewife (*Pomolobus pseudoharengus* [Wilson]) in Seneca Lake, New York." PhD diss., Cornell University, 1934.

Osgood, Herbert L. *The American Colonies in the Seventeenth Century.* Vol. 1, *The Chartered Colonies: Beginnings of Self-Government.* New York: Macmillan Company, 1904.

Otis, James. *Mary of Plymouth: A Story of the Pilgrim Settlement.* New York: American Book Company, 1910.

Our County and Its People: A Descriptive and Biographical Record of Bristol County, Massachusetts. Boston: Boston History Company, 1890.

Pairpoint, Alfred J. *Rambles in America, Past and Present.* Boston: Alfred Mudge & Son, Printers, 1891.

"Pearl Essence: Its History, Chemistry and Technology." Appendix II in the *Report of the U.S. Commissioner of Fisheries for 1925.* Bureau of Fisheries Document No. 989.

Perlmutter, A. *Guide to Marine Fishes.* New York: New York University Press, 1961.

Plumb, Albert Hale. *When Mayflowers Blossom: A Romance of Plymouth's First Years.* New York: Fleming H. Revell Company, 1914.

Pory, John. Letter of John Pory to the Earl of Southhampton, January 13, 1622–23. Later included in James, Sydney V., Jr., ed. *Three Visitors to Early Plymouth: Letters About the Pilgrim Settlement in New England during Its First Seven Years.* Plymouth, MA: Plimoth Plantation Inc., 1963.

Prince, Thomas. *A Chronological History of New England, in the Form of Annals.* Boston: Cummings, Hilliard, and Company, 1826.

Pumphrey, Margaret B. *Pilgrim Stories.* Chicago: Rand McNally & Company, 1910.

Reback, K.E., P.D. Brady, K.D. McLaughlin and C.G. Milliken. *A Survey of Anadromous Fish Passage in Coastal Massachusetts.* Part 1, *Southeastern Massachusetts.* Massachusetts Division of Marine Fisheries Technical Report TR-15. Boston: Massachusetts Division of Marine Fisheries, Department of Fisheries and Game, May 2004.

Report of the Commissioner for the Year Ending June 30, 1898. Part 24. Washington, D.C.: U.S. Commission of Fish and Fisheries, 1899.

Report of the Commission on Waterways and Public Lands on the Water Resources of the Commonwealth of Massachusetts, Their Conservation and Utilization Together with the Report of the Engineer of the Commission, 1918. Boston: Commonwealth of Massachusetts, 1918.

Robbins, Maurice. "Historical Approach to Titicut." *Bulletin of the Massachusetts Archaeological Society* 11, no. 3 (April 1950).

———. *Wapanucket: A Report of the Archaeological Investigations Realized by the Members of the Cohannet Chapter of the Massachusetts Archaeological Society on the North Shore of Assawompsett Pond, Middleboro, Massachusetts.* Attleboro, MA: Trustees of the Massachusetts Archaeological Society Inc., 1980.

Romaine, Mertie. *History of the Town of Middleboro, Massachusetts.* Middleborough, MA: Town of Middleborough, 1969.

Rostlund, Erhard. "The Evidence for the Use of Fish as Fertilizer in Aboriginal North America." *American Journal of Geography* 56, no. 5 (1957): 222–28.

Rounsefell, G.A., and L.D. Stringer. "Restoration and Management of the New England Alewife Fisheries with Special Reference to Maine." *Trans. American Fish. Society* 73 (1945): 394–424.

Russell, Howard. *Indian New England Before the Mayflower.* Hanover, NH: New England University Press, 1980.

———. *A Long Deep Furrow: Three Centuries of Farming in New England.* Hanover, NH: New England University Press, 1976.

Ryder, John A. *Report, U.S. Comm. of Fish. (1885).* Washington, D.C.: U.S. Commission of Fish and Fisheries, 1887.

Sanderson, J.M. *The Complete Cook.* Philadelphia: J.B. Lippincott & Company, 1864.

Sanford, Enoch, Reverend. *History of Raynham, Mass., from Its First Settlement to the Present Time.* Providence, RI: Hammond, Angell & Company, Printers, 1870.

Scott, W.B. *Freshwater Fishes of Eastern Canada.* Toronto: University of Toronto Press, 1955.

Scott, W.B., and E.J. Crossman. *Freshwater Fishes of Canada.* Bulletin 184. Fisheries Research Board of Canada, 1973.

Shurtleff, Nathaniel, ed. *Records of the Colony of New Plymouth in New England.* Vol. 4, *Court Orders 1661–1668.* Boston: Press of William White, 1855.

———. *Records of the Colony of New Plymouth in New England: Judicial Acts, 1636–1692.* Boston: Press of William White, 1857.

Smith, John. *Advertisements for the Inexperienced Planters of New England, or Anywhere.* London: John Haviland, 1631.

Snow, Dean R. *The Archaeology of New England.* New York: Academic Press, 1980.

Snyder, Randal J., and Todd M. Hennessey. "Cold Tolerance and Homeoviscous Adaptation in Freshwater Alewives (*Alosa pseudoharengus*)." *Fish Physiology and Biochemistry* 29, no. 2 (May 2003): 117–26.

Speth, John, and Katherine Spielman. "Energy Source, Protein Metabolism, and Hunter-Gatherer Subsistence Strategies." *Journal of Anthropological Archaeology* 2 (1983): 1–31.

Spotila, James R., et al. "Temperature Requirements of Fishes from Eastern Lake Erie and the Upper Niagara River: A Review of the Literature." *Environmental Biology of Fishes* 4, no. 3 (August 1979): 281–307.

Thomas, Peter A. "Contrastive Subsistence Strategies and Land Use as Factors for Understanding Indian-White Relations in New England." *Ethnohistory* 23, no. 1 (Winter 1976): 1–18.

Thoreau, Henry David. *A Week on the Concord and Merrimack Rivers.* Boston: James Munroe and Company, 1849.

Trautman, M. *The Fishes of Ohio*. Columbus: Ohio State University Press (Waverly Press), 1957.

Tressler, Donald K., and James McW. Lemon. *Marine Products of Commerce: Their Acquisition, Handling, Biological Aspects, and the Science and Technology of Their Preparation and Preservation*. New York: Reinhold Publishing Company, 1951.

Ursin, M.J. *A Guide to the Fishes of the Temperate Atlantic Coast*. New York: E.P. Dutton, 1972.

Vogel, Virgil J. "The Blackout of Native American Cultural Achievements." *American Indian Quarterly* 11, no. 1 (Winter 1987): 11–35.

Watertown Records Comprising the First and Second Books of Town Proceedings with the Lands, Grants, and Possessions, Also the Proprietors' Book and the First Book and Supplement of Births, Deaths and Marriages. Watertown, MA: Watertown Historical Society, 1894.

Watts, Douglas. *Alewife: A Documentary History of the Alewife in Maine and Massachusetts*. Augusta, ME: Poquanticut Press, 2012.

Weeden, William B. *Economic and Social History of New England, 1620–1789*. Boston: Houghton, Mifflin and Company, 1890.

Weston Thomas. *History of the Town of Middleboro, Massachusetts*. Boston: Houghton, Mifflin and Company, 1906.

Williams, Roger. *A Key into the Language of America, or, An Help to the Language of the Natives in That Part of America, Called New England*. 6th ed. Bedford, MA: Applewood Books, n.d. Originally published, London: Gregory Dexter, 1643.

Winifred, Lady Howard of Glossop. *Journal of a Tour in the United States, Canada and Mexico*. London: Sampson Low, Marston & Company, Limited, 1897.

Winthrop, John. *The History of New England from 1630 to 1649*. Boston: Phelps and Farnham, 1825.

Wood, William. *New England's Prospect*. London: Thomas Cotes, 1634. Reprinted, Boston: Publications of the Prince Society, 1865.

UNPUBLISHED SOURCES

Copeland, Jennie. "Taunton Alewives." Unpublished ms, Taunton Public Library, document no. 557.

"A List of Herrings to Be…to the Poor on Account of the Town of Middleboro' for the Year 1820 Appropriated by the Committee Appointed for that Purpose by Said Town, at Their Annual March Meeting." Unpublished ms, Middleborough Historical Association, mss no. 429. Gift of Elmer Drew.

Lutins, Allen. "Prehistoric Fishweirs in North America." Master's thesis, Master of Arts in Anthropology, Graduate School of the State University of New York at Binghamton, May 1992.

Maddigan, James F., Jr. Papers, including schematic maps of the Star Mills site and general notes on the herring in Middleborough. In the possession of Michael J. Maddigan.

Maddigan, Michael J. "Japan Works: A History of the George H. Shaw Company Site, East Grove Street, Middleborough." Prepared for the Middleborough Historical Commission, February 1996.

"Middleboro Town Treasurers Book, July 18th 1769." Middleborough Historical Association.

Rice, George S., and George E. Evans. "New Bedford Water Works. Further Water Supply. Plan Showing Proposed Overflow from Great Quittacas Pond into Pocksha Pond," June 9, 1897. Plymouth County Registry of Deeds, Plan Book 1, page 188 (serial number 988).

MAPS

Middleborough, Massachusetts

Middleboro, Mass. New York: Sanborn Map & Publishing Company, Limited, August 1885.

Middleboro, Plymouth Co., Mass. New York: Sanborn-Perris Map Company, Limited, May 1891.

Middleboro, Plymouth Co., Mass. New York: Sanborn-Perris Map Company, Limited, June 1896.

Middleboro, Plymouth County, Mass. New York: Sanborn-Perris Map Company, Limited, April 1901.

Taunton, Massachusetts

Insurance Maps of Taunton, Massachusetts. 1898. New York: Sanborn-Perris Map Company, Limited, 1898.

Insurance Maps of Taunton, Massachusetts, Including Town of Raynham. 1937. New York: Sanborn Map Company, 1937.

Insurance Maps of Taunton, Massachusetts, Including Town of Raynham. 1937. New York: Sanborn Map Company, 1937, amended June 1948.

Taunton, Bristol County, Massachusetts. Nov. 1893. New York: Sanborn-Perris Map Company, Limited, 1893.

Taunton, Massachusetts. July 1888. New York: Sanborn Map & Publishing Company, Limited, 1888.

DIRECTORIES

Taunton Directory, 1870–71; Containing the City Record, the Names of the Citizens, and a Business Directory with Other Useful Information. Boston: Sampson, Davenport & Company, 1870–71.

The Taunton Directory, 1869; Containing the City Record, the Names of the Citizens and a Business Directory with Other Useful Information. Boston: Sampson, Davenport & Company, 1869.

The Taunton Directory, 1925. Boston: Sampson & Murdock Company, 1924.

INDEX

ABOUT THE AUTHOR

Michael J. Maddigan has been involved in the field of local history and historic preservation for more than thirty years. He has written extensively on the history of Middleborough and Lakeville, Massachusetts, and is the author of several books on local history, including *South Middleborough: A History*, previously published by The History Press. Other works include *Elysian Fields: A History of the Rock Cemetery* (2007), *Images of America: Middleborough* (2009), *An Illustrated History of the King Philip Tavern* (2010), *Star Mill: History and Architecture* (2012) and *Representatives of the Great Cause:*  *Middleborough Servicemen & Their Letters from the First World War* (2013). He has contributed articles to numerous publications, and his work currently appears in the *Middleboro Gazette* as the popular local history column "Recollecting Nemasket." Maddigan owns a small publishing press and website under the Recollecting Nemasket name, both of which are devoted to popularizing local history. He is currently at work on separate histories of the Brockton Fair and the Bridgewater State Farm.